The Design, Performance, and Analysis of Slug Tests

The Design, Performance, and Analysis of Slug Tests

James J. Butler, Jr.
Kansas Geological Survey
The University of Kansas

Library of Congress Cataloging-in-Publication Data

Butler, James J.
The design, performance, and analysis of slug tests / James J. Butler, Jr.
p. cm.
Includes bibliographical references and index.
ISBN 1-56670-230-5 (alk. paper)
1. Groundwater flow--Measurement. 2. Aquifers--Measurement. 3. Wells--Testing. I. Title.
GB1197.7.B88 1997
551.49--dc21 97-24597
CIP

Direct all inquiries to CRC Press LLC, 2000 Corporate Blvd., N.W., Boca Raton, Florida 33431.

Lewis Publishers is an imprint of CRC Press LLC

International Standard Book Number 1-56670-230-5
Library of Congress Card Number 97-24597
Printed in the United States of America 1 2 3 4 5 6 7 8 9 0
Printed on acid-free paper

Preface

"What do I do with data like these?"

About six years ago, that simple question started me down the path that led to this book. My colleagues and I frequently found ourselves working with slug-test data that could not be readily analyzed with the commonly used methods. When we turned to the literature for assistance, we discovered few answers for those confronted with non-ideal data. In fact, other than the useful, but not widely available, literature reviews of Chirlin (1990) and Boak (1991), we could not find a single general reference to which field practitioners could go for answers to practical questions about slug tests. Although the need for such a reference was widely recognized, no one seemed prepared to invest the time required to put the necessary material together. Finally, in a moment of frustration, I decided to take a stab at it.

This book is primarily designed to be a practical reference text. However, in the guise of a practical reference, this book also addresses some fundamental problems currently limiting the utility of information obtained from slug tests. Most ground-water scientists and engineers would agree that without more attention to data-acquisition methodology, the considerable promise of numerical models will never be realized. The slug test can potentially provide very useful information about the transmissive and storage properties of a unit, and their variations in space, on a scale of relevance for a variety of modeling investigations. Unfortunately, however, much of the data currently being obtained from slug tests are, and often rightfully so, viewed with considerable skepticism. This book is my attempt to place the slug test on sounder theoretical and procedural grounds with the goal of improving both the actual and perceived quality of information obtained with the technique. Although this book should certainly not be considered as the final word on the topic, I hope it can serve as a useful reference for the near future.

I am the only author of this book, and thus solely responsible for its contents. However, the research that underlies this effort should not be construed as the product of my labors alone. Many individuals contributed to various aspects of this work. Of particular note are my long-time colleagues at the Kansas Geological Survey, Carl McElwee and John Healey. Carl and I have worked together for a number of years on slug tests, as is evident by our coauthorship of several publications on the technique. Our cooperation in this and related work has been an extremely positive influence on my career. John Healey, our field hydrogeologist par excellence here at the Kansas Geological Survey, has been the source of much advice and assistance on the practical aspects of the methodology. John's background in the drilling industry was of particular assistance in preparation of Chapter 2. I also want to acknowledge the contributions of Geoff Bohling of the Kansas Geological Survey and Gil Zemansky of Compass Environmental. Geoff was the primary author of the Suprpump analysis package, variants of which I used to perform many of the analyses discussed in this book. Sunday morning jogs with Gil have been the source of invaluable information about current practices in the consulting industry.

Many students at the University of Kansas provided field support for this work. These include Wenzhi Liu, Xiaosong Jiang, Tianming Chu, Yahya Yilmaz, Kristen Stanford, and Zafar Hyder. I would particularly like to cite the contribution of Wenzhi Liu. The careful reader will note that many of the tests presented as examples in this book were performed between late September and mid-November, a period during which the weather in Kansas can be particularly fickle. As a result, Wenzhi and I often ended up working under conditions that were considerably less than ideal. His good humor and ability to withstand the cold, wind, and dust without complaint were greatly appreciated.

I would also like to acknowledge the contributions of several individuals who kindly shared with me the products of their work. Kevin Cole of the University of Nebraska spent all-too-many hours generating the results presented in Tables 5.5, 6.3, and 6.4. Abraham Grader of Pennsylvania State University also went beyond the call of duty to generate simulation results that were an important contribution to Chapter 10. Frank Spane of Pacific Northwest Laboratory provided a copy of the DERIV program and valuable advice drawn from his extensive experience with various slug-test methods. Srikanta Mishra of Intera and Chayan Chakrabarty of Golder and Associates were both quite helpful in providing unpublished/in-press manuscripts.

I would also like to thank several individuals whose contributions were of a less-tangible nature. Bruce Thomson of the University of New Mexico provided friendship and excellent restaurant recommendations for the "Duke City" in the early stages of this project. Vitaly Zlotnik of the University of Nebraska was quite helpful with pithy comments and as a font of Russian aphorisms, the meaning of which I honestly never understand (fortunately, "ah, yes" seems to be the appropriate response in most cases). Rex Buchanan, Rich Sleezer, and cohorts provided editorial comments and comic relief in the final stages of this project. Finally, I would like to acknowledge the invaluable contributions of three individuals in the administration of the Kansas Geological Survey: Lee Gerhard, Larry Brady, and Don Whittemore. As a result of their efforts, the Kansas Geological Survey has certainly been an exciting place to pursue research in applied hydrogeology.

Although the above individuals all made important contributions to this effort, the most significant contributions were those of my family. This book could not have been written without the wholehearted support of my wife, Yun, and our children, Bill and Mei. Yun displayed much forbearance in allowing me the all-too-many nights and weekends of work that were needed to complete the book, while also serving as expert draftsperson and as all-purpose spiritual advisor. I greatly look forward in the coming months to spending much more time with Yun and the gang, and much less time with this computer.

The Author

Jim Butler is an associate scientist with the Geohydrology Section of the Kansas Geological Survey. He holds a B.S. in Geology from the College of William and Mary, and a M.S. and Ph.D. in Applied Hydrogeology from Stanford University. His primary research interest is in the development of field methodology for site characterization. Additional professional activities include teaching in the Department of Geology of the University of Kansas, acting as a consultant to federal agencies and private industry, and serving as an associate editor of the journal *Water Resources Research*. Previously, he has held positions as a visiting scientist in the Geohydrology Department of Sandia National Laboratory and as a graduate researcher in the Institute of Geology of the State Seismological Bureau in Beijing, China.

Table of Contents

1 Introduction

In virtually all groundwater investigations, one needs to have an estimate of the transmissive nature of the subsurface material that is the focus of study. In hydrogeology, the transmissive nature of the media is characterized by the parameter termed hydraulic conductivity or, in its fluid-independent form, intrinsic permeability. A large number of experimental techniques have been developed over the years to provide estimates of the hydraulic conductivity of subsurface material. These techniques range from laboratory-based permeameter or grain-size analyses to large-scale multiwell pumping tests. In the last two decades, a field technique for the estimation of hydraulic conductivity *in situ* known as the slug test has become increasingly popular, especially among scientists and engineers working at sites of suspected groundwater contamination. It is no exaggeration to say that literally tens of thousands of slug tests are presently being performed each year in the United States alone. Despite the heavy utilization of this technique in environmental applications, relatively little has been written about the practical aspects of the methodology. Many articles can be found in the technical literature on theoretical models for the analysis of slug-test data; yet there has been relatively little published on how to actually apply the technique in practice. Given the prevalence of the technique and the economic magnitude of the decisions that may be based on its results, there is clearly a pressing need for a text to which the field investigator can refer for answers to questions concerning the design, performance, and analysis of slug tests. The purpose of this book is to fill that need.

THE SLUG TEST — WHAT IS IT?

The slug test is a deceptively simple approach in practice. It essentially consists of measuring the recovery of head in a well after a near-instantaneous change in head at that well (a nearby observation well can also be used in certain situations). Figure 1.1 is a hypothetical cross-section through a confined formation in the vicinity of a monitoring well that illustrates the major features of a slug test. In the standard configuration, a slug test begins with a sudden change in water level in a well. This can be done, for example, by rapidly introducing a solid object (hence the term "slug") or equivalent volume of water into the well (or removing the same), causing an abrupt increase (or decrease) in water level. Following this sudden change, the water level in the well returns to static conditions as water moves out of the well (as in Figure 1.1) or into it (when change was a decrease in water level) in response to the gradient imposed by the sudden change in head. An example record of water-level changes during a slug test is given in Figure 1.2. These head changes through time, which are termed the response data, can be used to estimate the hydraulic

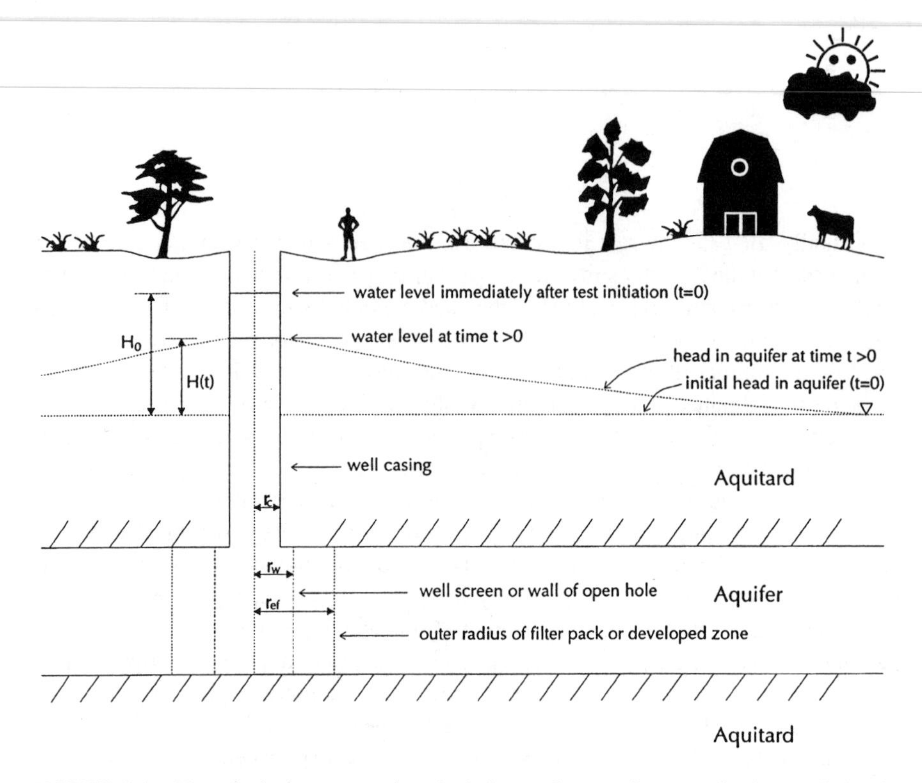

FIGURE 1.1 Hypothetical cross section depicting a slug test in a monitoring well that is fully screened across a confined aquifer (slug test initiated at time t = 0; figure not to scale).

conductivity of the formation through comparisons with theoretical models of test responses. In certain conditions, the slug test can also be used to obtain an estimate of the formation's ability to release or accept water into storage. This storage capability of the media is characterized in hydrogeology by the parameter designated as specific storage.

The parameter estimates obtained from slug tests can be used for a variety of purposes. At sites of suspected groundwater contamination, test estimates can be used to predict the subsurface movement of a contaminant, to design remediation schemes, and to plan multiwell pumping tests for obtaining more information about the large-scale hydraulic behavior of the subsurface units of interest. In water supply investigations, slug-test estimates are primarily utilized for the design of large-scale pumping tests and for the assessment of the effectiveness of well-development activities at observation wells. In near-surface agricultural applications, parameter estimates obtained with the slug test, which is termed the auger-hole or piezometer method in the agricultural and soils literature, can be used to design drainage systems for lowering shallow water tables. In petroleum and coalbed-methane applications, parameter estimates from slug tests and the closely related drillstem test are primarily used to help assess the potential for economic exploitation of a particular petroleum- or methane-producing horizon.

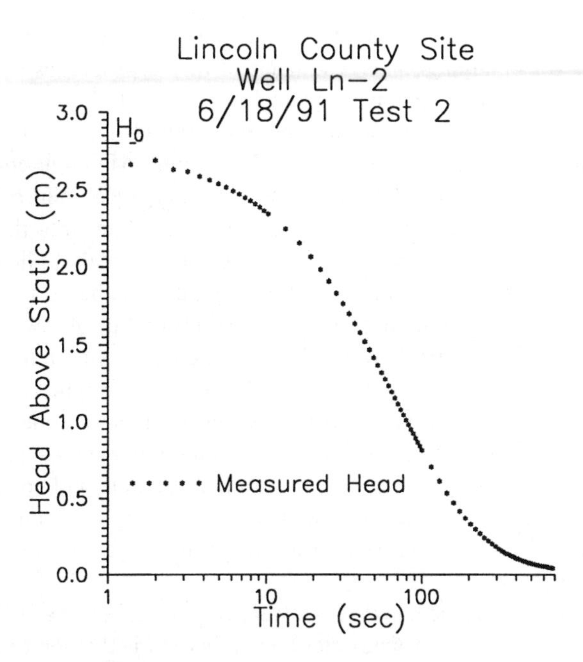

FIGURE 1.2 Head above static conditions vs. log time plot of a slug test performed in well Ln-2 at the Lincoln County, Kansas monitoring site (H_0 denotes magnitude of initial displacement).

WHY IS IT SO PREVALENT?

The slug test has become such a frequently used field technique as a result of its considerable logistical and economic advantages over alternative approaches. The most important of these advantages are:

1. Low cost — both in terms of manpower and equipment, the slug test is considerably less expensive than alternative approaches. A program of slug tests can be performed by one, or, at most, two people using a pressure transducer, data logger, and minor amounts of auxiliary equipment. When the cost of the equipment is spread over a large number of tests, the cost per test is extremely low;
2. Simplicity — as described in the previous section, the slug test is an extremely simple procedure. One initiates a test by a variety of means and then just measures the changes in head through time. Other than the possibility of having to clean equipment before moving to the next well, little else is required in the field;
3. Relatively rapid — the duration of a slug test is quite short in formations that would be classified as aquifers. In less-permeable formations, the test duration can be made relatively short through appropriate test design (e.g., decreasing the effective casing radius);

4. Very useful in tight formations — the slug test may be one of the best options for obtaining *in situ* estimates of media properties in formations of low hydraulic conductivity. In such so-called "tight" units, it may not be practical to perform constant-rate pumping tests because of the difficulty of maintaining a very low discharge rate. Although constant-head injection tests are often performed in the geotechnical industry, the logistics of the approach and the need to introduce water into the formation make this technique less attractive for environmental applications. Historically, laboratory testing of core samples has been a widely used approach for obtaining information on the properties of low-conductivity media. This technique, however, has become less common in recent years because of the concern that core samples may not provide information on a large enough scale to detect the existence of preferential flow paths, which can be very important conduits for fluid movement in such settings. The difficulty of obtaining an "undisturbed" sample and concerns about possible differences between the vertical and horizontal components of hydraulic conductivity have further limited the use of core-based approaches;
5. No water needs to be handled — a very important advantage of slug tests at sites of suspected groundwater contamination is that the technique can be configured so that no water is actually removed from or added to the well during the course of a test. This can be done by initiating a test through the addition or removal of a solid slug from the water column, the pressurization-depressurization of the air column in the well, etc.;
6. Provides information on spatial variations in hydraulic properties — a program of slug tests can be designed so that information about variations in the transmissive and storage properties of a formation can be obtained at a scale of relevance for contaminant transport investigations. Conventional pumping tests will provide large-scale volumetric averages of hydraulic properties, which may be of limited use in transport investigations. By performing a series of slug tests at discrete vertical intervals within individual wells and/or single tests in relatively closely spaced wells, important information can be obtained about the vertical and horizontal variations in hydraulic properties at a site;
7. Perceived straightforward analysis — the analysis of response data from slug tests is generally perceived to be extremely straightforward. Most analysis methods involve fitting straight lines or type curves to plots of field data. The boundary effects that may make the analysis of data from large-scale pumping tests quite involved generally have little to no impact on the response data from slug tests.

BUT SKEPTICISM ABOUNDS ...

Despite the heavy usage by the environmental industry, the slug test is viewed quite skeptically by many groundwater scientists and engineers. The origin of this skepticism is the discrepancy that is often observed between hydraulic conductivity estimates obtained from slug tests and estimates obtained from other elements of the field investigation (e.g., geologic and geophysical logs, pumping tests, etc.). Although spatial variability and the different scales at which the various estimates were obtained can explain a portion of the observed discrepancy, I believe that there are two other factors that are primarily responsible for this situation. First, well-development activities are often quite minimal at monitoring wells, the primary type of wells in which slug tests are performed. The result is that slug-test data are heavily impacted by drilling-induced disturbances and products of biochemical action. Countless field examples demonstrate the significant impact of insufficient well development on slug-test results. Unfortunately, such situations may very well be the norm, and not the exception. Second, the simplicity of the technique seems to breed a certain degree of casualness among those involved in the performance and analysis of slug tests. The result is that many of the assumptions underlying conventional analysis techniques are not upheld, introducing a considerable degree of error into the final parameter estimates. Fortunately, greater attention to details of well construction and development, coupled with the application of more care to all aspects of the design, performance and analysis phases, can greatly improve this situation. However, as will be emphasized throughout this book, the effects of incomplete well development may be difficult to avoid. Thus, the hydraulic-conductivity estimate obtained from a slug test should virtually always be viewed as a lower bound on the hydraulic conductivity of the formation in the vicinity of the well.

PURPOSE OF THIS BOOK

The major purpose of this book is to provide a series of practical guidelines for the design, performance, and analysis of slug tests that should enable reasonable parameter estimates to be obtained from slug tests on a consistent basis. Three critical themes will be emphasized throughout this book:

1. Importance of well development — slug tests are extremely sensitive to near-well disturbances; so, it is no exaggeration to say that the success of a program of slug tests critically depends on the effectiveness of well-development activities. Repeat slug tests and preliminary screening analyses will be the primary approaches recommended here for assessment of the effectiveness of well-development activities;
2. Importance of test design — a program of slug tests must be designed to assess whether the assumptions underlying conventional analysis methodology are viable for a particular set of tests. In this book, repeat slug tests will be the primary method proposed for this assessment;
3. Importance of appropriate analysis procedures — the analyst must assess the hydrogeologic setting of the test well and use the most appropriate data-analysis procedure for that setting. Preliminary screening of the response data and comparison of the expected and measured values for the initial displacement will be proposed here as the primary means for defining the most appropriate procedure to employ for a particular hydrogeologic setting.

The importance of these themes will be demonstrated throughout the book using a series of field examples. The vast majority of these examples will be drawn from field studies done by the author while at the Kansas Geological Survey.

In keeping with its role as a reference text, the target audience for this book is very broad, ranging from the practicing professional to the academically oriented investigator. An attempt was made to provide a thorough discussion of all practical issues involved in the performance, design, and analysis of slug tests. For example, each analysis method is clearly outlined in a step-by-step manner, after which the procedure is illustrated with a field example and all major practical issues related to the field application of that technique are discussed. For the more theoretically minded reader, the mathematical models underlying all major techniques are presented, thus allowing the assumptions incorporated in the various analysis methods to be better understood. The ultimate objective of the presentation is to help the reader explore a given topic to virtually any depth that is desired.

OUTLINE

The heart of this book consists of the following 12 chapters (Chapters 2 to 13). Each chapter is designed to be a relatively self-contained unit, so that the reader can refer to a particular section without necessarily needing to read the earlier chapters in detail. The major points of a chapter are summarized in the form of a series of practical guidelines that are given at the conclusion of each chapter or, in the case of the analysis methods, presented in a separate summary chapter (Chapter 12).

Chapter 2 will focus on the design of a series of slug tests, the all-too-often neglected phase of a test program. Details of well construction and development pertinent to slug tests will be discussed, with a special emphasis placed on approaches for assessing the sufficiency of well development activities. A lengthy presentation on the use of repeat slug tests to verify the appropriateness of the assumptions underlying the conventional analysis methodology will also be given.

Chapter 3 will focus on issues associated with the performance of slug tests, the most practical aspects of a test program. The primary types of equipment that are used for the measurement and storage of head data will be described. The most common methods for initiating a slug test will then be presented, and the strengths and weaknesses of each method will be highlighted. A particular emphasis will be placed on assessing each method with respect to the relative speed of test initiation and the potential to obtain an accurate estimate of the head change initiating a test.

Chapter 4 will focus on the pre-analysis processing of response data, a critical step for preparing data for formal analysis and for assessing the appropriateness of the assumptions invoked by conventional analysis methods. A special emphasis will be placed on the processing of data collected with pressure transducers.

Chapters 5 through 12 will focus on the techniques employed for the analysis of head data from slug tests. All major methods for the analysis of slug tests in confined and unconfined formations will be described. Examples will be heavily utilized to illustrate the appropriate way a particular approach should be applied. A special emphasis will be placed on preliminary screening analyses, the results of which will determine the specific approach to employ in the actual analysis. A series of flow charts will be presented in Chapter 12 to aid in the selection of appropriate analysis methodology.

Chapter 13 will briefly summarize the major themes of the book, emphasizing the most critical elements of the practical guidelines. Appendices following this chapter will provide a short discussion of available software for the analysis of slug test data and will define notation used in the text. The book will conclude with a list of all references cited in the text.

A SHORT WORD ON TERMINOLOGY ...

Over the last 40 years, a fair amount of terminology has been developed with respect to slug tests. Unfortunately, certain aspects of this terminology have led to some degree of confusion and misunderstanding. Two aspects are worth noting here. First, there has been an effort to differentiate between tests that are initiated by a sudden rise or a sudden drop in the head in a well, i.e., tests in which the direction of the slug-induced flow (into/out of the well) differs. For tests initiated by a sudden rise in head, the terms falling-head, slug, slug-in, and injection tests have been most commonly employed in the literature. For tests initiated by a sudden drop in head, the terms rising-head, bail-down, bailer, slug-out, and withdrawal tests have been most commonly used. The term response test has been used for both situations. In this book, the term slug test will be used for all tests in which the focus of interest is the response to a near-instantaneous change in head at a well. If there is a need to differentiate between tests on the basis of the direction of the slug-induced flow, the modifiers rising-head and falling-head will be employed. Second, the head change initiating a slug test has been called the slug, the initial displacement, H_0, and the slug-induced disturbance, among other things. In this book, the terms initial displacement and H_0 will primarily be used to designate this initial head change.

One final issue of semantics concerns what to call the individuals who are primarily responsible for the planning, performance, and analysis of a program of slug tests. The most appropriate designation, "groundwater scientists and engineers," is a bit too lengthy for repetitive use; so, more succinct terminology must be employed. Thus, the terms "hydrogeologist" and "hydrologist" will be used interchangeably in this book to designate the group of scientists and engineers of a multitude of backgrounds who are charged with the task of carrying out/overseeing a program of slug tests.

2 The Design of Slug Tests

CHAPTER OVERVIEW

The degree of success of a program of slug tests is primarily a function of a series of design decisions, some of which are made long before the actual tests are performed in the field. The emphasis of this chapter will be on the major elements involved in the design phase of a test program. The discussion will primarily focus on the three most critical areas of the design phase: (1) well construction, (2) well development, and (3) verification of conventional theory. The chapter will conclude with the definition of a series of practical guidelines for the design of a program of slug tests.

WELL CONSTRUCTION

A well is the hydrogeologist's window on the subsurface. The specific approaches used to drill, install, and develop a well will largely determine the clarity of the view through that particular "window". In this section, the drilling and installation aspects of the construction process will be briefly examined. Well development will be the focus of the following section.

The hydrogeologist has a wide range of drilling techniques to choose from when considering the well-construction phase of a project. Driscoll (1986) summarizes most of the major methods employed in groundwater investigations, while Aller et al. (1989) provide a series of matrices that can aid in the selection of an appropriate drilling technique. The particular approach used at a site will depend on, among other things, the hydrogeologic conditions at the site and the specific purpose for which that well is to be used. The most common well type at sites of potential groundwater contamination is the monitoring well. Since one of the major purposes of a monitoring well is to obtain representative samples of the groundwater at that particular location, drilling techniques that limit the use of substances that might affect later chemical analyses of the groundwater are often favored over other approaches. Thus, for wells that are to be drilled in consolidated rock, air-rotary techniques are often selected over techniques using mud or water as the drilling fluid, while augering or driving methods are preferred in unconsolidated formations. Although driving using cable-tool techniques can be employed in both consolidated and unconsolidated formations, this approach is not commonly used for monitoring wells because it is much slower than other options.

Regardless of which drilling technology is employed, a considerable amount of drilling debris will be generated in the construction process. This debris, which could include remnant drilling fluids, fine material created or mobilized by the drilling

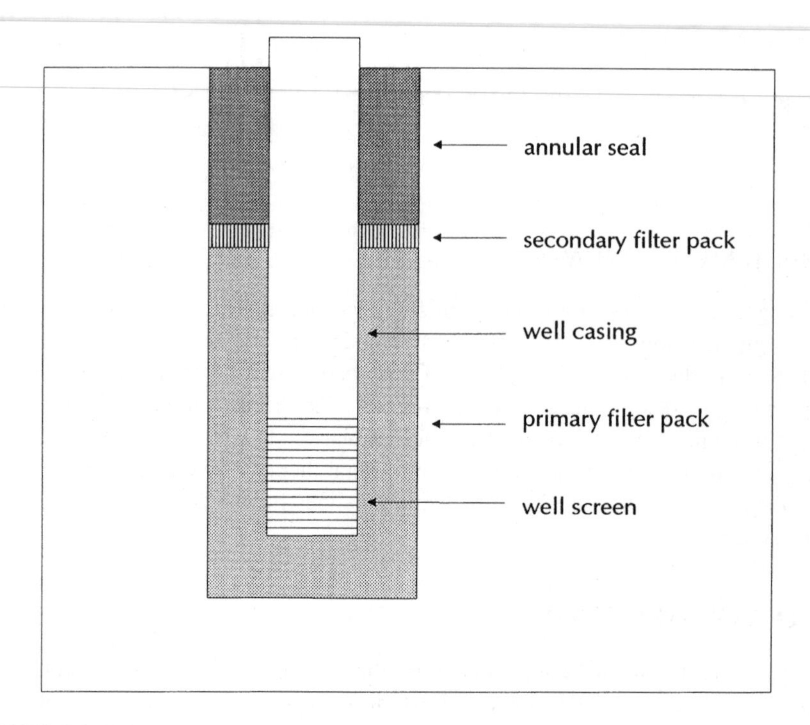

FIGURE 2.1 Hypothetical cross section depicting the major elements of a monitoring well (figure not to scale).

process, compacted sediments, clay smears, etc. will be concentrated in the near-well portions of the formation. The specific nature of the debris will depend on the particular technique used to drill the well. For example, drilling debris will primarily consist of very fine pulverized material in wells that have been constructed using air-rotary techniques, while compacted sediments and clay smears will be the primary components in driven wells. Given the difficulties that may be encountered in trying to remove this debris from the formation, some consideration should be given to use of techniques that minimize its creation. Driving methods, whether traditional cable-tool approaches or the higher-frequency techniques used in shallow applications, are widely considered to be the best options for minimizing drilling-induced disturbances (e.g., Morin et al., 1988).

Once a borehole has been drilled to the target depth, the next step is to actually install the well. There are four major elements of a well that must be considered in the installation process, each of which has some bearing on slug tests that are performed at that well. These elements are, as shown in Figure 2.1, the well casing, the well screen, the filter pack, and the annular seal. Aller et al. (1989) and Nielsen and Schalla (1991) provide detailed discussions of these features from a monitoring-well perspective, while Driscoll (1986) and Roscoe Moss Company (1990) focus on water-supply wells. The implications of each of these elements for slug tests will be briefly reviewed here.

Although most discussions of the well casing revolve around the structural integrity and chemical resistance of the casing material, these aspects are of limited relevance for slug tests. Instead, the radius of the casing is of most importance, as it will control both the duration of the test and the type of equipment that can be used in the well (see Chapters 3 and 7). Moreover, in small-diameter wells sited in media of very high hydraulic conductivity, the velocity in the casing may be great enough to produce additional head losses that can complicate the analysis of response data (see Chapter 8). McElwee and Butler (1996) have shown that wells greater than 0.05 m in diameter may be required to minimize these effects in highly conductive media.

The major issues with respect to the well screen are the size of the openings, the percent of openings relative to the total surface area, the length of the screen, and the nature of the inner surface of the screen. The size of the openings (slots) should be determined by the characteristics of the formation in which the well is screened and/or the artificial filter pack. Slot size is most critical in wells without an artificial filter pack, as inappropriate sizing can greatly complicate well development efforts and, therefore, potentially introduce errors into the hydraulic-conductivity estimate obtained from a slug test. In screens with a relatively small percent of open surface area, such as might be found when the slots have been made by hand or a downhole casing perforator, convergent flow to the sparsely distributed slots may produce additional head losses that can complicate the analysis of test data. However, most commercially available machined-slotted or continuous-wrapped screens will have a sufficiently large enough open area to minimize such effects. The length of the well screen is primarily an issue with respect to the efficiency of well development, as the longer the screen the greater the potential to have significant portions of the interval that have been virtually untouched by the development process. The inner surface of the screen is only of concern for continuous wrapped varieties, where the vertical rods upon which the screen is wrapped may limit the options for well development. In particular, these rods may make it difficult to isolate and develop discrete portions of the screened interval.

The filter or gravel pack is the name given to the material that is placed in the annular space between the inner diameter of the borehole and the outer diameter of the well screen. If the material is added from the surface, it is termed an artificial filter pack. If the filter pack consists of formation material that collapses against the screen when the support provided by the drill pipe or temporary casing is removed, it is termed a natural filter pack. In either case, the major purpose of a filter pack is to stabilize the formation by decreasing the potential for movement of fine material into the well. In stable formations (i.e., those prone not to collapse when support is removed), another important function of a filter pack is to provide a support for the overlying annular seal. Regardless of the type of filter pack that is employed, it will usually form a zone of higher hydraulic conductivity immediately outside the screen. As is discussed in the following section, this zone of higher conductivity may hinder the effectiveness of well development activities.

The annular seal is the name given to material of very low hydraulic conductivity that is emplaced above the filter pack in the space between the outer diameter of the casing and the inner diameter of the borehole. The primary purpose of this seal is to prevent vertical movement of water along the annulus created by drilling. Such vertical flow could introduce anomalous effects into a slug test and confound attempts to analyze the response data. The seal material normally consists of a bentonite- or cement-based product, usually emplaced in the form of a grout. If the material is introduced in a grout form, a secondary filter pack of fine sand and/or a thin layer of bentonite pellets is placed on top of the primary filter pack as illustrated in Figure 2.1. This additional layer(s) is designed to prevent movement of the grout into the filter pack opposite the screened interval.

Regardless of which of the methods described in Chapters 5 through 11 are used to analyze the response data, estimates must be provided for the effective (as opposed to nominal) dimensions of the well casing and screen. Since most of the issues concerning these effective quantities are related to well development, discussion of this topic will be deferred to the following section.

WELL DEVELOPMENT

As emphasized in the preceding section, a considerable amount of debris will be concentrated in the near-well portions of the formation as a result of the drilling process. Well development is the term given to a class of post-drilling procedures, one of whose objectives is to remove drilling debris from the near-well portions of the formation adjacent to the screened (open) interval. It is no exaggeration to say that well development is the single most important aspect of a program of slug tests. Unfortunately, however, well development is all too often a neglected component of a field investigation. The result is that the parameter estimates obtained from slug tests may have a rather tenuous connection to reality.

Two examples can be used to demonstrate the critical importance of well development for slug tests. Figure 2.2 displays normalized head data (measured deviations from static normalized by the magnitude of the initial displacement) from a series of slug tests performed on the same day at a site in Stafford County, Kansas. Response data can be compared using T_0, which, as will be explained more fully in Chapter 5, is the basic time lag (i.e., time at which a normalized head of 0.37 is obtained [Hvorslev, 1951]) and inversely proportional to hydraulic conductivity. In this series of tests, T_0, and thus the estimated hydraulic conductivity, varied by a factor of 2.3. Apparently, some fine material is being mobilized during these tests and is moving in a manner that produces large changes in hydraulic conductivity between successive tests. The magnitude of the changes that can occur as a result of the movement of near-well fine material was further demonstrated by a later test performed at this same well which yielded a T_0 value of under 88 s. It is important to point out that these tests were performed after this well had been extensively developed in late 1994. A slug test performed at this same well approximately 14 years earlier, shortly after installation and after relatively little development, yielded a T_0 estimate of over 20,000 s. This over two order of magnitude difference in T_0 values from tests at the same well is a graphic illustration of the critical need for proper well development prior to the performance of slug tests. Although the development that was done in late 1994 greatly improved near-well conditions, the slug

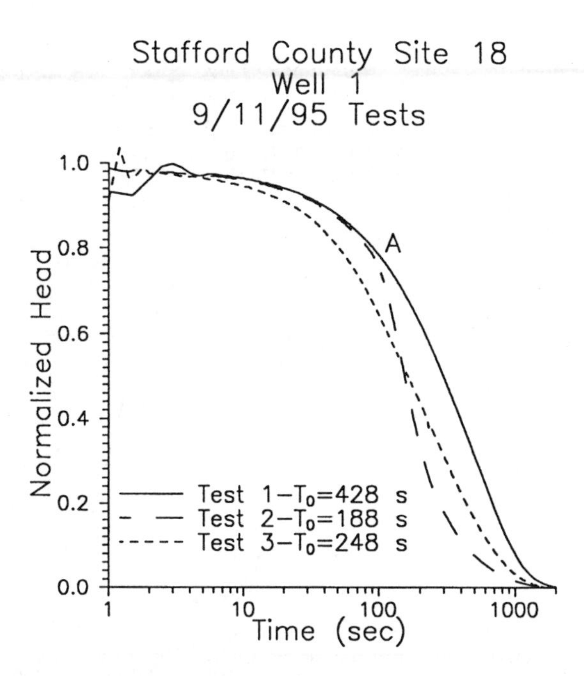

FIGURE 2.2 Normalized head ($H(t)/H_0$, where H(t) is measured deviation from static and H_0 is magnitude of the initial displacement) vs. log time plot of a series of slug tests performed in well 1 at monitoring site 18 in Stafford County, Kansas (T_0 and A defined in text).

tests performed following that effort, a subset of which are shown in Figure 2.2, indicate that development is clearly not complete.

The series of tests shown in Figure 2.3 serves as a second demonstration of the importance of well development for slug tests. In this test series, T_0 varied by close to two orders of magnitude. Tests labelled "rh" (rising head) were configured so that the direction of the slug-induced flow was into the well, while the test labelled "fh" (falling head) was configured so that the direction of flow was out of the well. Mobilization of near-well fine material was again considered the most likely cause for the differences seen between the two rising-head tests. The large contrast between the rising- and falling-head tests, however, is undoubtedly due to a different mechanism. In this case, the explanation considered most likely is that products of biochemical action on or in the immediate vicinity of the well screen are acting like a check valve, causing flow into the well to meet much less resistance than flow in the opposite direction (Butler and Healey, 1995).

These two field examples clearly show that slug tests can be heavily influenced by altered, near-well conditions. This altered, near-well zone is commonly termed the well skin, and can have a hydraulic conductivity that may be lower (low-K skin) or higher (high-K skin) than the formation itself. As will be discussed in detail in Chapter 9, low-K skins will have a much more dramatic impact on slug-test response data than high-K skins, and thus are the primary focus of interest here.

Well skins can also be classified with respect to their stability during a series of slug tests. If the properties of the near-well zone significantly change during a set

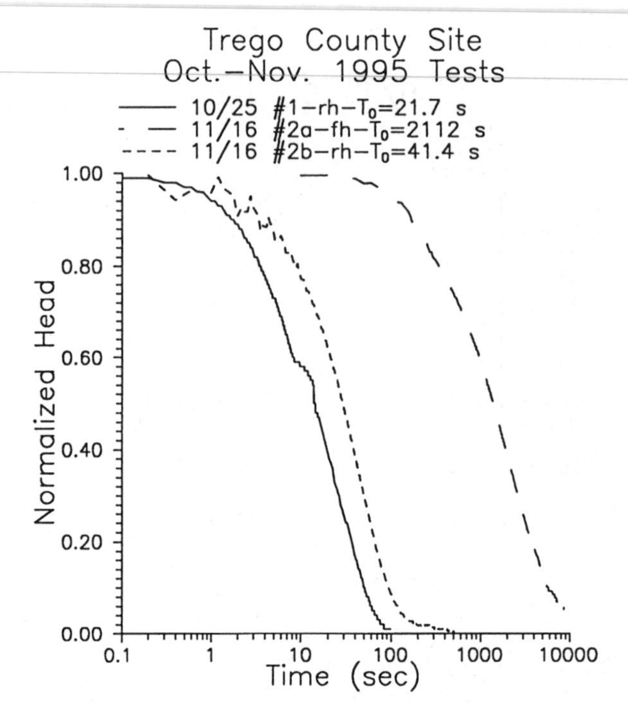

FIGURE 2.3 Normalized head ($H(t)/H_0$) vs. log time plot of a series of slug tests performed in a monitoring well in Trego County, Kansas (T_0, rh, and fh defined in text).

of tests as a result of, for example, the mobilization and movement of fine material, the skin is considered to be dynamic or evolving in nature according to the terminology proposed by Butler et al. (1996). In such a case, sizable changes in T_0 values are seen between repeat slug tests performed at a well (e.g., Figure 2.2). These changes often appear to be largely random in nature and not dependent on the direction of the slug-induced flow or the magnitude of the initial displacement. If the skin remains essentially unchanged between tests, it is considered to be static in nature. For example, in the series of slug tests performed at a Trego County, Kansas monitoring well, a subset of which are given in Figure 2.3, there was as much as a factor of two difference between T_0 estimates from repeat tests that had the same flow direction. This difference, however, was quite small when compared to the nearly two order of magnitude difference consistently seen between rising- and falling-head tests. Although the skin was altered somewhat between tests with the same flow direction, the essential characteristic of the skin did not change. Thus, this can be considered an example of a predominantly static skin.

The primary goal of well development activities for slug tests is to remove drilling-related debris and/or products of human-induced biochemical action from the near-well portions of the formation. Unfortunately, given the rather limited attention that well development receives in most site investigations, the conditions illustrated in Figures 2.2 and 2.3 may very well be the norm, and not the exception. Butler and Healey (1998) speculate that the commonly observed difference between hydraulic-conductivity estimates obtained from slug tests and those obtained from pumping tests (slug-test estimates are usually less than estimates from pumping tests in the same formation) is primarily a product of well skins that are of lower hydraulic conductivity than the formation itself. Although theoretical models for slug tests in wells with skins exist, these models are of limited use in practice for a variety of reasons that will be discussed in Chapter 9. The result is that it may be virtually impossible to remove the effects of a low-K skin in the analysis phase of a slug test. Well development is therefore the most effective approach for diminishing the impact of a low-K skin on the parameter estimates obtained from slug tests.

Well drillers historically have employed a great variety of methods for the development of wells that are to be used for water-supply purposes. These range from simple pumping and surging to the introduction of various fluids (disinfectants, acids, surfactants, etc.) to the use of downhole explosive devices (Driscoll, 1986; Roscoe Moss Company, 1990). The focus of this overview of well development, however, will be on methods that can be used in monitoring wells at sites of potential groundwater contamination. Since one of the primary functions of a monitoring well is to obtain representative samples of the groundwater at a particular location, only a relatively narrow range of development options can be considered. For example, methods that introduce fluids to the well (even, in some cases, potable water) are usually not employed because of their potential influence on later chemical analyses. Moreover, at sites of suspected groundwater contamination, development approaches that require the removal of considerable quantities of water are usually avoided because of the cost involved with the handling and disposal of potentially contaminated water. The result is that the primary techniques used for development of monitoring wells are removal of limited quantities of water via pumping, bailing, or airlifting, and surging via a variety of means. Aller et al. (1989), Kill (1990), and ASTM (1996), among others, provide detailed descriptions of the various methods used in the development of monitoring wells, while Roscoe Moss Company (1990) describes the derivation and application of mathematical models of various development methods. A few general aspects of the development process will be highlighted here.

Although a series of recommended practices for monitoring well development have been promulgated in the literature (e.g., ASTM, 1996), fiscal constraints have, in all too many cases, limited development activities to simply pumping a well until a relatively clear stream of water is produced. The position of the pump is often not changed and very limited surging is done. As has been demonstrated repeatedly in the field, such an approach can be a very ineffective method of development. The reasons for this can be readily shown using Figure 2.4, which displays a series of hypothetical cross sections through a formation in the vicinity of a monitoring well. Figure 2.4A depicts typical conditions shortly after well installation has been completed. Drilling debris essentially lines the entire zone opposite the filter pack. If the well is stressed from a location above the screen, the result will be a situation similar to that shown in Figure 2.4B, where development is limited to a few channels through the drilling debris, most of which will be in the vicinity of the screened interval. However, even in those channels, the position of which will undoubtedly be strongly correlated with the location of the more permeable intervals of the

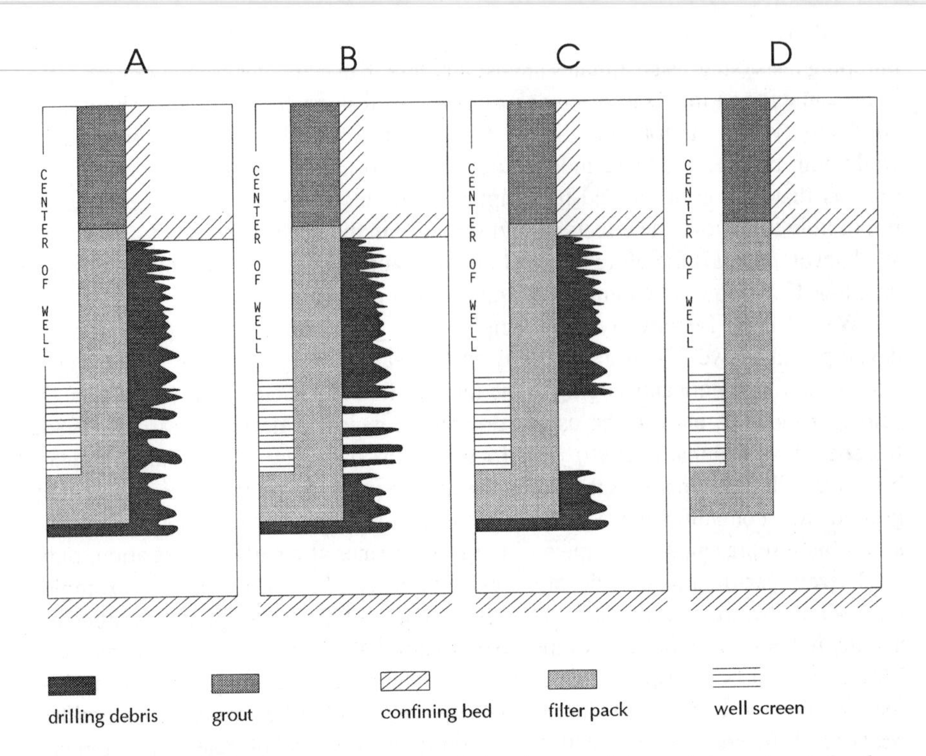

FIGURE 2.4 Hypothetical cross section depicting four possible conditions of well development for a partially penetrating well (figure not to scale).

formation in the vicinity of the screen, the development may be quite limited. If development procedures can be configured so that the stress imposed by the development activity is concentrated in discrete intervals of the screen (e.g., through use of straddle packers or a surge block in the screened interval), then it may be possible to reach a situation that is reasonably close to that of Figure 2.4C. Unfortunately, however, this level of development is not commonly achieved in monitoring wells. Although often assumed for the analysis of slug-test response data, the situation shown in Figure 2.4D is virtually impossible to obtain in practice, given the standard well installation and development procedures employed in groundwater investigations.

Regardless of which development method is used, the most critical issue is how to assess if more development is necessary. The decision on when development has been sufficient will depend on the particular application for which a well is to be used. For example, if the primary purpose of a well is to obtain a representative sample of the groundwater for chemical analysis, a measure of the turbidity of the water is the most common criterion for assessing the sufficiency of well development (Wilson, 1995). Slug tests are usually considered a secondary application at most monitoring wells; so, development activities are rarely judged in a slug-test context. However, since slug tests are extremely sensitive to near-well conditions, a test program must be designed to assess whether insufficient development is affecting

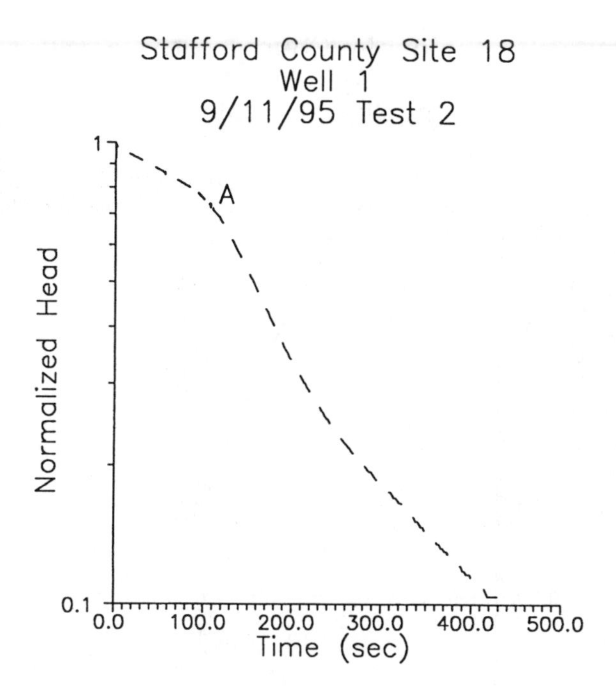

FIGURE 2.5 Logarithm of normalized head ($H(t)/H_0$) vs. time plot of a slug test performed in well 1 at monitoring site 18 in Stafford County, Kansas (A defined in text).

the response data. The identification of a low-K skin at a well is the clearest indication that further development activities are needed if reliable parameter estimates are to be obtained from slug tests at that well.

Three procedures can be used for recognizing the presence of a low-K skin with a single slug test. Probably the most common approach is to compare the hydraulic-conductivity estimate obtained from a slug test with other information collected as part of the site investigation. This other information could include geologic and geophysical logs, core samples, estimates from nearby pumping tests, etc. If the slug-test estimate appears to be much lower than what would be expected from these other data, then the existence of a low-K skin can be considered as a likely explanation for the discrepancy. However, given the range of hydraulic conductivity variations possible in natural systems, the difference must be quite large to justify the low-K skin interpretation.

A second approach is to recognize dynamic-skin effects from a plot of the test data. In some cases, dynamic-skin effects may be so pronounced that there is a clear break on the data plot. For example, the break at point A for test 2 on Figure 2.2, which is more clearly seen in the log head vs. time format of Figure 2.5, is a clear indication of a dynamic skin when the well is screened below the water table. In most cases, however, dynamic-skin effects will be difficult to recognize on the basis of a single test.

The most effective approach for identifying the presence of a low-K skin with data from a single slug test is a comparison of test responses with theoretical models.

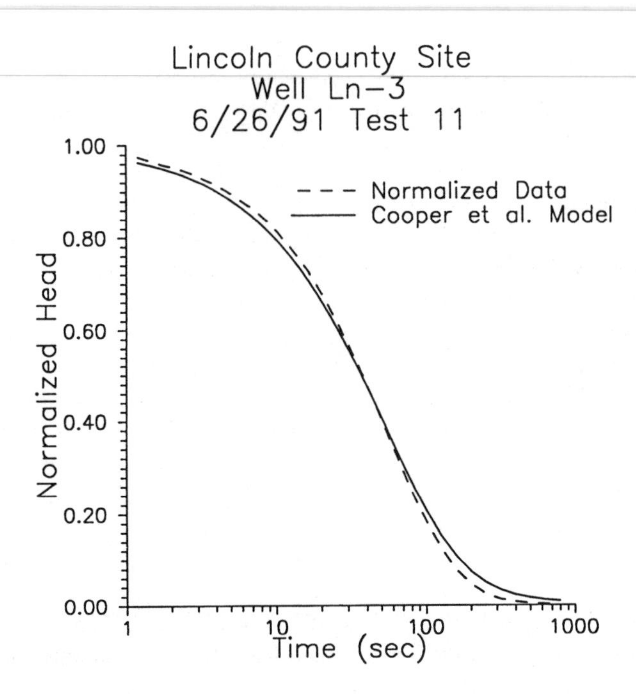

FIGURE 2.6 Normalized head ($H(t)/H_0$) vs. log time plot of a slug test performed in well Ln-3 at a monitoring site in Lincoln County, Kansas and a Cooper et al. type curve ($\alpha = 1.94 \times 10^{-4}$).

When a skin is present, a systematic deviation between the response data and the best-fit type curve from a theoretical model for slug tests in homogeneous formations (e.g., Cooper et al., 1967; Hyder et al., 1994) may be observed if the specific storage estimate is constrained to physically plausible values. Figure 2.6 shows such a deviation for the case of a skin estimated to be approximately one order of magnitude less permeable than the formation (Butler and Liu, 1997). A larger deviation is seen as the contrast between the conductivity of the skin and that of the formation increases (e.g., Figure 10 of Hyder et al., 1994). If the specific storage estimate is not constrained to the range of physical plausibility, a theoretical type curve will often closely fit the response data. In this case, however, the physically implausible specific storage estimate (very low in the case of a low-K skin) will be a clear indication that a skin is affecting the response data. Unfortunately, the absence of a systematic deviation or a physically implausible specific storage estimate does not necessarily imply the absence of a low-K skin, since such effects may not be observed if the slug-induced flow is constrained to a few well-developed zones through the drilling debris (e.g., Figure 2.4B).

As emphasized in the following section, the performance of repeat slug tests at a well enables the appropriateness of conventional slug-test theory to be assessed at that well. Butler et al. (1996) show how data from a program of repeat tests can be used to identify the dynamic (Figure 2.2) and directional-dependent (Figure 2.3) character of many low-K skins. At least three slug tests should be performed at each well for this purpose using initial displacements (H_0) in the range of 0.2 to 2.0 m. In order to assess the directional dependence of test responses, one test should be configured so that the direction of the slug-induced flow is opposite that of the other tests. Results similar to those presented in Figures 2.2 and 2.3 should be considered unambiguous indications that further development is necessary at that well.

Unfortunately, regardless of which method is used, it can be very difficult to detect conditions similar to those shown in Figure 2.4B. This will be especially true in formations with a significant degree of vertical anisotropy, where the vertical component of hydraulic conductivity is considerably less than the horizontal component. In such a situation, the anisotropy may result in the test data mimicking the response of a fully penetrating well in which the effective length of the well screen is the thickness of the channels through the drilling debris. For this reason, close agreement between T_0 values from repeat slug tests and reasonable fits between test data and theoretical type curves with physically plausible parameters can only be considered a partial demonstration of the sufficiency of well development. Thus, as emphasized throughout this book, the hydraulic-conductivity estimate obtained from a slug test should always be viewed as a lower bound on the hydraulic conductivity of the formation in the vicinity of the well.

Given the difficulty of effectively developing the entire screened interval of wells that have been installed in the conventional manner and of diminishing the impact of a low-K skin on parameter estimates obtained from an analysis of the response data, it may be worthwhile to consider the use of procedures in the post-drilling phase of the well-construction process that diminish the potential for the creation of conditions similar to those shown in Figure 2.4B. A few possible approaches are briefly summarized here.

In boreholes drilled in unstable formations (i.e., those prone to collapse when support is removed), a critical consideration is whether or not to use an artificial filter pack. Although an artificial filter pack can prevent the pumpage of excessive amounts of fine material, it can also serve as a conduit for vertical flow, thereby diminishing the effectiveness of development activities. The alternative, a natural filter pack, will usually have a vertical component of hydraulic conductivity much closer to that of the formation and, therefore, have less potential for significant vertical flow within the filter pack. If the primary goal of well development is to diminish the potential for the creation of conditions similar to those shown in Figure 2.4B, natural filter packs should be used in place of artificial filter packs whenever possible. If an artificial filter pack is deemed necessary, some thought should be given to diminishing the potential for vertical flow by altering the vertical component of the hydraulic conductivity of the pack material. This could be done, for example, by periodically placing very thin layers of lower-permeability material (i.e., finer sands) in the filter pack through use of either prepacked screens or equipment that allows accurate emplacement of thin layers in the well annulus from the surface. These finer-grained layers should be quite thin, so that there is little impact on horizontal flow, and should be sized to be compatible with the well screen, i.e., the average grain size should be large enough to prevent movement of significant amounts of this material into the well.

In stable formations, an attempt should be made to develop the well prior to the emplacement of the filter pack. If at all possible, the well should be developed using a straddle-packer system in which the packers are inflated against the formation. In most situations, the use of such a system should enable the well to be developed quite thoroughly. If development activity is not possible prior to the emplacement of the filter pack, a filter pack with periodic thin layers of lower-permeability material should again be considered as a means to improve the efficacy of later well development activities.

The effectiveness of well development activities is often a function of the length of the well screen and, to a lesser extent, that of the filter pack. Development in wells with short screen lengths and with filter packs that terminate a relatively short distance above the top of the screen may have a higher potential for success because the portions of the formation that are opposite the screen and filter pack may be relatively homogeneous. However, given the variability in hydraulic conductivity that has been observed over short vertical distances in formations that appear to be quite homogeneous in character (e.g., Smith, 1981; Sudicky, 1986), there is no guarantee of near-homogeneous conditions with respect to hydraulic conductivity when screens of very short length are employed. Thus, even in wells with short screens, development techniques that stress discrete intervals of the screened section are highly recommended.

If one strongly suspects that the response data at a given well are being significantly affected by a low-K skin, then additional well development is clearly needed before reliable parameter estimates can be obtained from slug tests at that well. Unfortunately, however, it may often be very difficult to remove the remnant drilling debris and/or biochemical products that comprise the low-K skin in a cost-effective manner. In such cases, virtually the only option for obtaining a reliable estimate of the hydraulic conductivity of the formation is to perform a short-term pumping test. The exact configuration of this test (i.e., discharge vs. injection, constant rate vs. constant head, etc.) will depend on specific site conditions. Butler (1990) has shown that a semilog Cooper-Jacob analysis (Cooper and Jacob, 1946; Kruseman and de Ridder, 1990) can often be used to remove the effects of a low-K skin from the hydraulic-conductivity estimate obtained from a constant-rate pumping test. In the case of a constant-head test, however, it may prove more difficult to remove the effects of a low-K skin from the conductivity estimate (Novakowski, 1993).

As mentioned at the end of the preceding section, all methods for the analysis of slug-test data require estimates of the effective dimensions of the well casing and screen (henceforth designated as the well-construction parameters). The effective length of the well screen is the well-construction parameter most likely to introduce error into the hydraulic-conductivity estimate obtained from a slug test. The nominal length of the screened interval and the length of the filter pack are the two most common quantities used for the effective screen length. Because the filter pack is usually considerably more permeable than the formation, many authors (e.g., Palmer and Paul, 1987; Butler et al., 1996) have recommended using the length of the filter pack for the effective screen length. However, this recommendation is based on the often unrealistic assumption that near-well portions of the formation have been well developed along the entire length of the filter pack (Figure 2.4D). The reality is undoubtedly something much closer to the conditions depicted in Figure 2.4B. Since it may be very difficult to quantify the thickness of the channels that have been developed through the drilling debris, Butler (1996) has recommended that the nominal length of the well screen be employed for the effective screen length and that the resulting hydraulic-conductivity estimate be considered as a lower bound on the conductivity of the formation. This recommendation is supported by the results of a series of numerical simulations that indicate that the nominal length of the well screen should be used for the effective screen length in virtually all situations (Butler, 1996).

The effective radius of the well screen is also required for the analysis of response data. The nominal radius of the well screen and the radius of the filter pack are the two quantities commonly used for this parameter. Again, because the filter pack is usually considerably more permeable than the formation, many authors (e.g., Palmer and Paul, 1987; Butler et al., 1996) have recommended using the radius of the filter pack for the effective screen radius. In this case, the recommendation seems in keeping with the realities of well installation and development. Butler (1996) presents the results of a series of simulations that show that when the permeability of the filter pack is much greater than that of the formation (a factor of two or above), the radius of the filter pack is the most appropriate estimate of the effective screen radius. Thus, the radius of the filter pack should be used in virtually all cases when an artificial filter pack is employed. In the case of a natural filter pack, the nominal radius of the well screen may be more appropriate if well development has been limited.

The final well-construction parameter required for the analysis of response data is the effective casing radius. For conventional slug tests performed in confined flow systems or in wells in unconfined flow systems screened a considerable distance below the water table, the effective casing radius should be the nominal casing radius. One situation where this would not be true would be that of entrapped air in the filter pack; a condition most likely to occur in formations of relatively low permeability where the filter pack may often be placed into a nearly dry hole. Keller and Van der Kamp (1992) have shown that the expansion/compression of entrapped air in the filter pack will produce responses that are analogous to those from tests in wells with larger casing radii. In wells with screens or filter packs that extend across the water table, use of the nominal casing radius may also no longer be appropriate. In this case, Bouwer (1989) recommends use of a modified casing radius that incorporates the porosity of the filter pack material (see Chapter 6). As will be discussed further in Chapter 3, a comparison of the theoretical initial displacement estimated from volumetric calculations based on the nominal casing radius with the initial displacement estimated from the test data is an excellent check on the appropriateness of the nominal casing radius. Close agreement between these two estimates is strong support for use of the nominal casing radius as the effective radius parameter. Note that in shut-in slug tests, which will be discussed extensively in Chapter 7, the effective casing radius is a function of the compressibility of the water and test equipment (Neuzil, 1982), and thus may be quite different from the nominal radius of the casing.

VERIFICATION OF CONVENTIONAL THEORY

After well development, the next most important aspect of a program of slug tests is the specific design for the planned series of tests. A series of slug tests must be designed to evaluate the validity of conventional slug-test theory at a given well. There is no point in analyzing response data with a particular theoretical model if there is a significant difference between the underlying assumptions of that model and the actual conditions at a site. Fortunately, a series of slug tests can be readily designed to assess the validity of many of the most critical assumptions of conventional theoretical models.

In wells screened in confined formations or in unconfined formations where neither the well screen nor the filter pack intersects the water table, conventional theory (e.g., Cooper et al., 1967; Hyder et al., 1994) maintains that response data from repeat slug tests should coincide when graphed in a normalized format (measured deviations from static normalized by the magnitude of the initial displacement (H_0)). Thus, if conventional theory is applicable, the normalized response data should be independent of the size of the initial displacement and the direction of the slug-induced flow. Butler et al. (1996) recommend that a minimum of three slug tests be performed at each well to assess the appropriateness of conventional theory at that well. The magnitude of the initial displacement should be varied by at least a factor of two, and the first and last tests of the series should have the same initial displacement. These procedures will, among other things, allow the effects of a dynamic skin to be readily separated from a reproducible dependence on H_0 that is characteristic of tests in media of very high hydraulic conductivity (see Chapter 8).

The direction of the slug-induced flow should be varied during a series of tests to identify a skin-related directional dependence. The primary direction of flow during a test series (i.e., the flow direction in the majority of the tests) should be from the formation into the well. Butler (1996) notes that flow into the formation can often lead to a progressive decrease in near-well hydraulic conductivity as a result of mobilized fine material being lodged deeper in the formation. Initiating the majority of tests so that the direction of the slug-induced flow is into the well (i.e., a rising-head test) may diminish the severity of that phenomenon.

If a well has been developed according to recommended procedures (e.g., ASTM, 1996), the normalized response data from a series of slug tests at that well coincide, and there is no indication of a low-K skin (e.g., a physically implausible specific storage estimate) from a preliminary analysis of the test data, then one can be confident that conventional slug-test theory is applicable at that well. For example, Figure 2.7 displays response data from a series of three tests performed at a monitoring site in Stafford County in south central Kansas. The coincidence of the normalized data indicates that test responses are independent of the size of the initial displacement, that dynamic skin effects are quite small, and that there is no skin-related directional dependence. Close agreement between the theoretical and measured values for the initial displacement indicates that the nominal casing radius should be the effective casing radius for these tests. In addition, Butler et al. (1993) show that a close match between test data and a theoretical type curve can be obtained using a physically plausible specific storage estimate. Thus, all evidence points to the viability of conventional theory at this well. The primary source of potential error in this case is that introduced by incomplete well development, which may have resulted in a situation similar to that shown in Figure 2.4B, i.e., the slug-induced flow has been constrained to a few channels through a static skin. Thus, the hydraulic-conductivity estimate obtained from these tests can only be considered as a lower bound on the hydraulic conductivity of the formation.

The mathematics underlying the conventional theoretical models used for the analysis of response data are linear in nature. However, this linear theory may not be appropriate in wells that are screened close to or across the water table. If the water table can be represented as a constant-head boundary (i.e., little to no change in position), then conventional theory is applicable. However, if the position of the water table markedly changes during a test, resulting in significant variations in the effective length of the screen through which water flows into/out of the well, then conventional theory is no longer applicable.

A series of slug tests in a well screened across or near the water table should be designed to assess which conceptualization of the water table is appropriate at that well. A minimum of three tests should be performed at each well. As with the procedures discussed earlier, the magnitude of the initial displacement should be varied by at least a factor of two between tests, and the first and last tests of the series should have the same initial displacement. In this case, a reproducible dependence on H_0 is either an indication of highly permeable media (non-Darcian flow effects) or of changes in the effective screen length. Varying the direction of the slug-induced flow between tests can help identify which of these possibilities is most likely. If the effective length of the screen changes during a test, then there should be a difference between rising- and falling-head tests at that well. Data from repeat tests with different flow directions should exhibit characteristics similar to the tests shown in Figure 2.8; the rising-head tests should be of longer duration and should have a pronounced concave-downward curvature, while falling-head tests should have the characteristic concave-upward curvature often seen in tests in confined formations. Note that in addition to an indication of the movement of the water table or of a low-K skin, a dependence on flow direction in a well screened across the water table may be an indication of changes in the effective casing radius. It is not difficult to envision a situation where the effective casing radius for a falling-head test (water level moving in screened section opposite unsaturated intervals) and that for a rising-head test (water level moving in screened section opposite saturated intervals) are different (Stanford et al., 1996). As is discussed in Chapter 3, a comparison of the theoretical and measured values for the initial displacement will clarify whether the effective casing radius is changing between tests. Given the strong dependence of test duration on the effective casing radius, changes in that parameter can result in tests of significantly different durations. In summary, a series of slug tests in a well screened across the water table must be carefully designed to elucidate the appropriate manner to represent the water table, whether a low-K skin is affecting test responses, whether non-Darcian flow effects are significantly influencing test data, and whether the effective casing radius depends on flow direction.

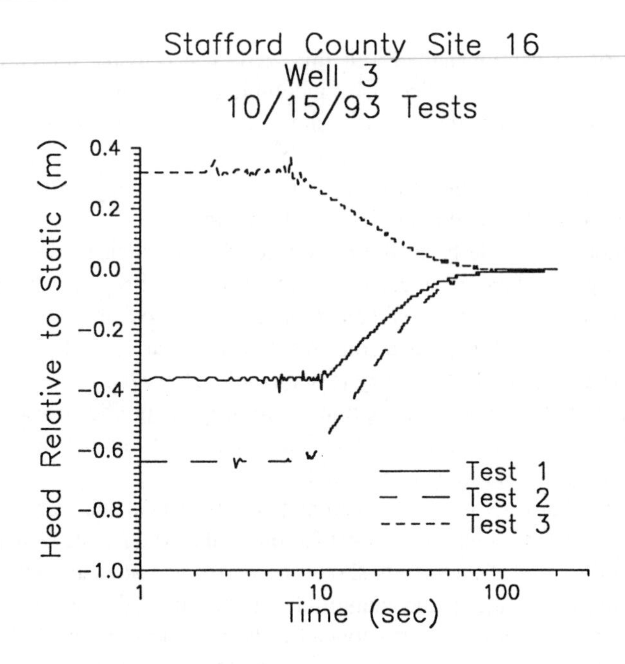

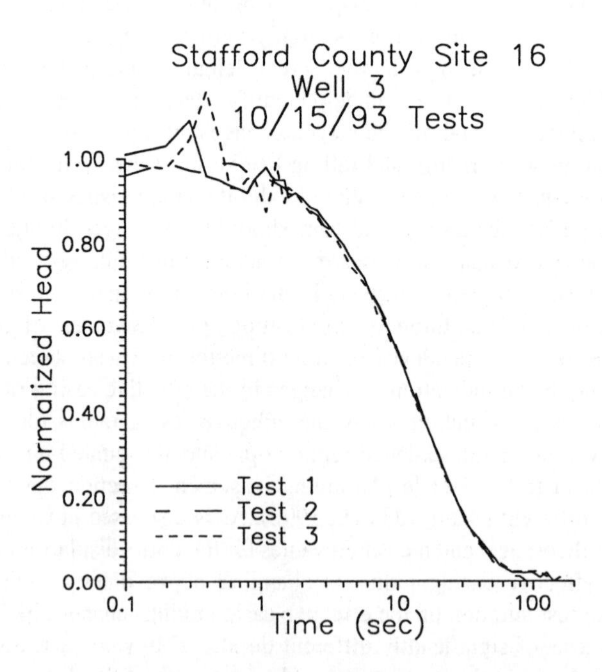

FIGURE 2.7 Plots from a series of slug tests performed in well 3 at monitoring site 16 in Stafford County, Kansas: (A) Head vs. log of time since initiation of data collection; (B) Normalized head ($H(t)/H_0$) vs. log of time since test initiation (packer used for test initiation).

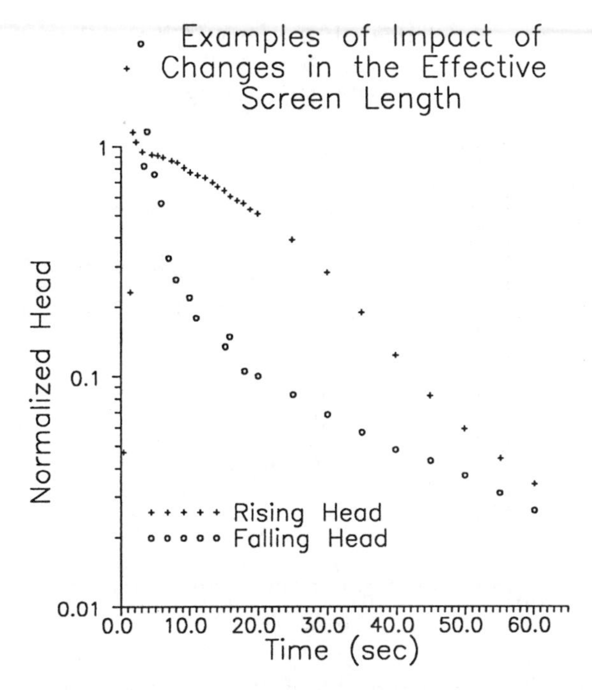

FIGURE 2.8 Logarithm of normalized head ($H(t)/H_0$) vs. time plot of a pair of rising- and falling-head tests performed in a well screened across the water table (response data from Dahl and Jones [1993]).

A field example can help illustrate several of the points made in the previous paragraph. Figure 2.9 displays normalized response data from a series of slug tests performed in a well screened across the water table in an unconsolidated sand and gravel aquifer. In this case, the tests were initiated by either the introduction or removal of a solid slug (see Chapter 3) from the well. The tests labelled sms (small slug) were performed using a solid slug that displaced 0.25 m of water, while the tests labelled lgs (large slug) were performed using a solid slug that displaced 0.56 m of water. Given the noise in the test data, all three rising-head (rh) tests can be considered to approximately coincide, indicating little dependence on the magnitude of the initial displacement. The falling-head (fh) tests, especially in the case of the large slug, are lagged in time in comparison to the rising-head tests, which is opposite of what would be expected if the effective screen length was changing during the course of a test. This lag is thought to be a result of a larger effective casing radius produced by small cavities along the well screen above the water table, most probably created during well installation (Stanford et al., 1996). The primary conclusion from this series of tests is that dynamic skin effects are relatively small, non-Darcian flow losses appear negligible, the water table can be conceptualized as a constant-head boundary, and that the effective casing radius is larger for the falling-head tests. The response data from these tests also indicate that flow to/from the unsaturated zone is playing a very minor role and can effectively be neglected, a result in keeping

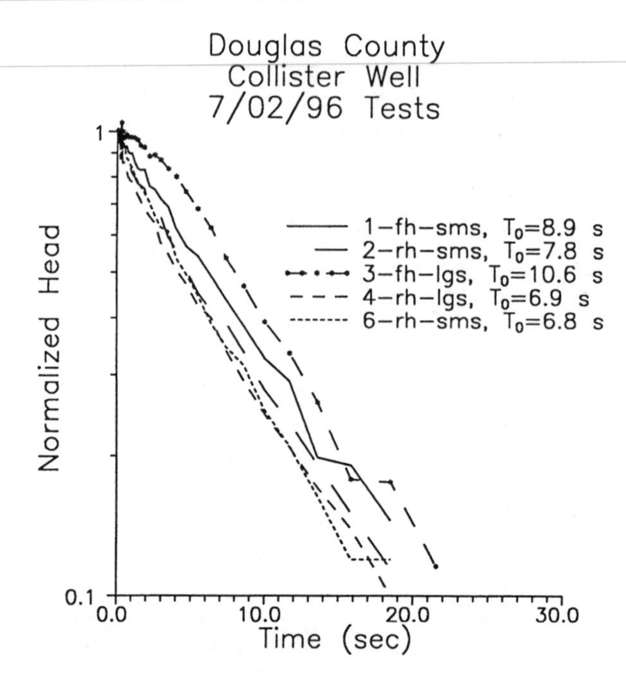

FIGURE 2.9 Logarithm of normalized head ($H(t)/H_0$) vs. time plot of a series of rising-head (rh) and falling-head (fh) tests performed in the Collister monitoring well in Douglas County, Kansas (sms and lgs defined in text, well screened across the water table).

with conventional theory on the role of the unsaturated zone during pumping tests (e.g., Kroszynski and Dagan, 1975).

DESIGN GUIDELINES

The following eight design guidelines can be extracted from the discussions of this chapter:

1. Well-drilling procedures that minimize the generation of drilling debris should be employed whenever possible. Driving based methods, such as cable-tool, pneumatic/hydraulic hammering, or rotosonic methods, are probably best in this regards;
2. Well-development activities should be directed at developing discrete intervals along the well screen. Development procedures that do not stress discrete portions of the well screen may prove rather ineffective, leaving substantial portions of the screened interval virtually untouched by development. Vertical flow within the filter pack can diminish the effectiveness of development efforts; so, some consideration should be given to use of post-installation procedures that may result in more complete development. These include development prior to emplacement of the filter pack in stable formations, use of specially constructed filter packs that decrease vertical flow, or use of natural filter packs in unstable formations;

3. The possibility of a low-K skin should be assessed by a preliminary analysis of the response data using a theoretical model for slug tests in homogeneous formations. A physically implausible specific storage estimate is strong evidence that a skin is affecting the response data;
4. The nominal screen length should be used for the effective screen length parameter in practically all cases;
5. The radius of the filter pack should be used for the effective screen radius parameter in wells with artificial filter packs, while the nominal screen radius may be a better choice for wells with natural filter packs if development has been limited;
6. The nominal radius of the well casing should normally be used for the effective casing radius in conventional slug tests. A comparison of the theoretical and measured values for the initial displacement will indicate the appropriateness of this recommendation for any particular test. The effective casing radius will be a function of the compressibility of water and test equipment in the case of a shut-in slug test;
7. Three or more slug tests should be performed at each well. Two or more different values for the initial displacement (varying by at least a factor of two) should be used in these tests. The first and last tests of the series should employ the same H_0, so that the effects of a dynamic skin can be separated from a reproducible dependence on the initial displacement. The direction of flow should also be varied between tests so that a skin-related directional dependence can be identified and, for the case of a well screened across or near the water table, the appropriate manner to represent the water table can be determined. These repeat tests should enable the effectiveness of well-development activities and the viability of conventional slug-test theory to be evaluated at each well;
8. The primary direction of flow during a series of slug tests should be from the formation into the well. Slug-induced flow from the well into the formation will often lead to decreases in hydraulic conductivity as a result of mobilized fine material being lodged deeper in the formation.

3 The Performance of Slug Tests

CHAPTER OVERVIEW

Once a well has been developed and a test plan has been devised, the next step is to actually perform a series of slug tests at that well. The emphasis of this chapter will be on the major elements involved in the performance of slug tests. The discussion will primarily focus on three areas: (1) the equipment used in the performance of slug tests, (2) the methods and considerations involved in initiating a test, and (3) the degree of head recovery needed prior to repeating a test at a particular well. The chapter will conclude with the definition of a series of practical guidelines for the performance of slug tests.

SLUG-TEST EQUIPMENT

Equipment used for the performance of slug tests can be grouped into three general categories: devices for initiating a test, devices for measuring changes in head during a test, and devices for storing the head measurements. Since a description of the equipment used for test initiation requires some discussion of the various initiation methods, consideration of that topic will be deferred to the section on test initiation. The focus of this section will be on measurement and storage devices.

MEASUREMENT DEVICES

The equipment used for measuring head changes during a slug test can be separated into devices that measure the depth to water from a datum at the top of the casing and devices that measure the pressure exerted by an overlying column of water. A brief description of the most commonly used devices in both classes is given here.

For decades, the standard method for measuring water levels in hydrogeologic field investigations was the chalked steel tape (depth to water calculated from the distance between a datum at the top of the casing and the point at which the water has washed the chalk off the tape). Although this technique can certainly be used for tests in formations of low hydraulic conductivity, early practitioners of slug tests quickly recognized that this approach would not allow rapid enough measurement of water levels in more permeable formations. Ferris and Knowles (1963) suggested the use of a popper (more commonly known as a plopper) device that consists of a cylindrical weight attached to a steel tape. The cylindrical weight has a concave lower end so that contact with the water produces a plopping or popping sound. The depth to water is then read directly from the steel tape. Although measurements can

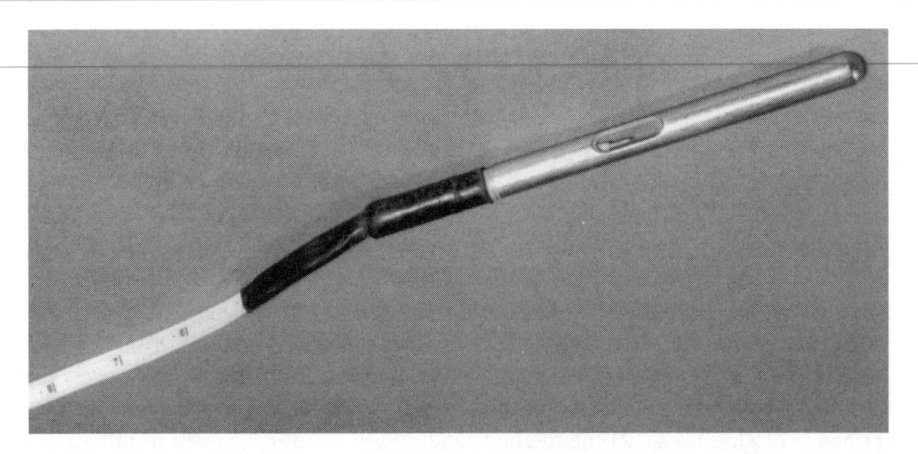

FIGURE 3.1 Lower end of an electric tape displaying the coaxial cable and the sensing element. Note the air gap between the two electrodes in the sensing element (diameter of the sensing element is approximately 1.3 cm.).

be taken at a much higher frequency than with a chalked tape, the need to actively "sound" each new level limits the general utility of the plopper for slug tests.

The electric tape has become a very commonly used tool for the measurement of water levels in slug tests in formations of moderate to low hydraulic conductivity. This device essentially consists of a two-wire coaxial cable that has electrodes separated only by an air gap at the lower end (Figure 3.1). The circuit is completed when both electrodes enter the water. A buzzer, light, or ammeter is used to indicate that the water level has been reached. Depth to water is read directly from a tape that is affixed to the cable. A reading is possible every 8 to 10 s in the hands of an experienced operator calling out measurements to a second individual recording depths and times. Very accurate and rapid depth measurements may be obtained by changing the position of the tape a set distance and then recording the time at which the water level passes by the electrodes. Electric tapes may provide spurious values in water of high electrical conductivity because the circuit can be completed by current moving through a thin water film that may be difficult to remove from the probe without pulling it from the well. Although the cable may stretch with time, cable stretch is probably not much of a concern for slug tests where water-level changes are the focus of interest.

The float recorder is also used in slug tests, although primarily only for tests in lower-permeability media. This device consists of a light weight (usually plastic) cylinder that floats on top the water column. A cable attached to the float and strung over a pulley arrangement at the top of the casing enables water-level changes to be recorded on a spring- or battery-driven strip chart. Conversion of the chart readings into a head vs. time record of sufficient resolution for the purposes of slug-test analyses may prove difficult. This problem, however, can be largely avoided by use of a depth encoder connected to a data logger. The depth encoder, which is attached to the axis of the pulley on which the float cable is strung, allows an electronic record of the float movement to be obtained.

FIGURE 3.2 Pressure transducer and cable. The pressure sensitive diaphragm is housed inside the black filter piece at the lower end of the transducer (diameter of the transducer is approximately 1.8 cm.).

Although electric tapes are frequently used for tests in formations of moderate to low hydraulic conductivity, the majority of slug tests are performed with sensors that measure the pressure exerted by the overlying column of water, the most common of which is the pressure transducer. A transducer is a device that converts a physical quantity (in this case, pressure) into an electrical signal. The most common type of pressure transducer used for slug tests is the semiconductor strain-gauge transducer (Figure 3.2). This device consists of a pressure-sensitive diaphragm on the back of which is bonded a series of semiconductor strain gauges. The front side of the diaphragm faces the water, while the back faces a chamber that is either at near-vacuum conditions or at atmospheric pressure (possible by means of a vent tube that runs the length of the cable). A pressure change in the water produces a minute movement of the diaphragm, which introduces strain to the gauges on the back side. This strain produces an electrical resistance change that is proportional to the applied pressure. This resistance change is apparent as a voltage difference, which is normally converted to a current (4 to 20 mA) output signal.

A second type of transducer is the vibrating wire transducer. In this case, a wire, which is attached to the back of the pressure sensitive diaphragm, is vibrated using plucking coils in the sensor. The degree of pressure-induced deflection of the diaphragm controls the frequency of the vibrating wire. The vibrating wire generates a current, the frequency of which can be converted to a pressure measurement. Although not common in groundwater investigations, this type of transducer is more frequently used in geotechnical applications.

The advantages of transducers are that the accuracy and resolution can be very good, measurements can be taken at a very high frequency, and the device allows unattended data collection (i.e., no one actually has to be on site) if paired with a data logger. The primary disadvantages of transducers are that they are complex pieces of equipment (i.e., can malfunction if not treated properly), they are relatively expensive, and they must be used with some sort of data-acquisition device. Transducers require periodic calibration in a controlled setting to compensate for long-term drift resulting from strain-hardening of the diaphragm, bonding deterioration, aging of circuitry, etc. It is very important to check transducer operation in the field prior to and at the end of each series of tests. A quick check can be performed by placing the transducer at a known depth in the well and comparing a depth-to-water estimate calculated from the transducer reading with a depth-to-water measurement obtained using an electric tape or similar device. Additional sources of error that can affect transducer operation are temperature shock and electrical interference. Prior to beginning a series of slug tests, a transducer should be allowed to equilibrate with the water in the well for at least 10 to 20 min (a greater length of time should be used when the air temperature is much different from the temperature of the groundwater). Failure to wait for temperature equilibration can introduce considerable error into the initial readings taken by the sensor. One also needs to be aware of error introduced by electrical interference from nearby power lines, etc. Use of a transducer with a current, in contrast to a voltage, output is recommended to avoid the possibility of such interference effects. Transducers can be obtained that measure absolute (back of diaphragm is at near-vacuum conditions) or gauge (relative to atmospheric) pressure. If a sensor measuring absolute pressure is used, atmospheric pressure changes occurring during a test must be removed from the pressure data prior to analysis.

Another approach for measuring the pressure exerted by the overlying column of water, which is primarily restricted to slug tests in low-permeability media, is the airline. In this case, a small-diameter tube extends from the top of the casing to a considerable distance below the top of the water column. The air pressure required to push all of the water out of the submerged portion of the tube is equal to the pressure exerted by the overlying column of water on the bottom of the tube. In slug tests in which water levels are falling through time, the tubing can be pressurized at the outset of the test and then decreases in tubing pressure can be recorded through time. Since the pressurization can be done with a small air compressor or tire pump, the technique is quite inexpensive. However, an air pressure gauge with high resolution and accuracy must be used to have measurements of reasonable quality. In slug tests in which water levels are rising through time, the tubing must be repressurized prior to each measurement to clear the tubing of water. A sophisticated form of the airline is the pneumatic pressure sensor. These sensors, however, are not common in groundwater applications.

DATA-STORAGE DEVICES

Head data collected during a slug test are initially stored in paper or electronic form. If transducers are not employed in the test, the data will most probably be entered into a field notebook. Although electronic equivalents of the field notebook are now available, recording of measurements with a pencil or pen (nonsoluble ink) in a field notebook is still the most common mode of data storage when transducers are not used.

If transducers are used in the test, some sort of electronic data acquisition and logging device must be employed. These devices can range from a data acquisition card/attachment for a personal computer to a stand-alone data logger. Essentially, all of these devices work in the same way. They receive an analog signal from a transducer, convert it to a digital form, process the digital data (i.e., convert the current or voltage output to physical units of pressure or head), and then store the results in an electronically retrievable format. The primary considerations for these devices are the maximum rate at which they can acquire and manipulate data, the resolution of the analog to digital (A-D) conversion, and their storage capacity. For slug tests in high-permeability media, a maximum acquisition rate of at least five measurements per second (5 Hz) is needed. The resolution of the A-D conversion is expressed by the number of bits used in the device. A minimum of 12 bits (resolution is 0.02% of measurement) is required to obtain data of a reasonable quality. Capacity to take measurements in equal log-time increments or to adjust the sampling rate during the course of a test is needed to prevent the storage capacity of the device from being exceeded prior to test completion. Note that a stand-alone data logger is often preferred over a data acquisition card/attachment because of its durability with respect to the nonideal conditions often faced in field settings and its capacity for unattended data collection.

In all cases, regardless of which storage device is employed, it is imperative that multiple copies of the relevant pages from the field notebook and/or of the electronic data record be made as soon as possible. Dependence on a single copy of the data, whether it be in paper or electronic form, is not a sound practice.

TEST INITIATION

There are two critical considerations regarding the initiation of a slug test: (1) the head change should be introduced in a manner that can be considered near instantaneous relative to the formation response; and (2) an estimate of the magnitude of this initial head change (designated as the initial displacement) should be obtained.

The utility of a particular method for test initiation will depend on the relative speed with which the formation responds to the head change and on α, the dimensionless storage parameter of the well-formation configuration ($\alpha = (r_w^2 S_s b)/r_c^2$, where r_w and r_c are the screen and casing radii, respectively, S_s is the specific storage, and b is the screen length). In cases where α is relatively small (less than about 0.001), the relative speed of initiation is of less importance because the head response will closely approximate an exponential decay in time (i.e., a plot of the log of the deviation of head from static conditions vs. time will be near linear in form). Butler (1996) has shown that, even if the head change is introduced quite slowly relative to the formation response, it may still be possible to obtain a reasonable estimate of the hydraulic conductivity of the formation when α is relatively small. However, since the α value will not be known in advance and the ideal conditions necessary to obtain a reasonable conductivity estimate may not exist, the head change should always be introduced as rapidly as possible.

Peres et al. (1989) state that an estimate of the initial displacement is not needed for analysis of slug-test data. This claim, however, is only true for the ideal conditions they examine. Actually, a reasonable estimate of the initial displacement is important for two reasons: (1) an estimate is needed to compare the expected (determined from volumetric calculations and designated H_0^*) and measured (determined in the field immediately after initiation and designated H_0) values for the initial displacement; and (2) an estimate is needed for comparison of repeat tests performed with different H_0.

A comparison between the expected and measured value of the initial displacement is most critical in three particular hydrogeologic settings: (1) wells in low-permeability formations, (2) wells screened across the water table, and (3) wells in high-permeability formations. Each of these is briefly described in the following paragraphs.

In low-permeability formations, a H_0 value that is less than expected may be an indication of entrapped air in the filter pack. Keller and Van der Kamp (1992) have shown that entrapped air in the filter pack will significantly delay test responses because the entrapped air results in an effective casing radius that is considerably larger than the nominal casing radius. Failure to recognize and account for the presence of entrapped air can therefore lead to a significant underestimation of the hydraulic conductivity of the formation. A difference between the expected and measured values for the initial displacement may be the only indication that entrapped air is affecting test responses.

In wells that are screened across the water table, a H_0 value that is smaller than expected may also be an indication of an effective casing radius that is larger than the nominal casing radius. Bouwer and Rice (1976) and Bouwer (1989) speculate that rapid dewatering of the filter pack may result in the radius of the filter pack or developed zone, modified by the porosity of the material in that zone, being the most appropriate quantity for the effective casing radius in wells screened across the water table (see Chapter 6).

Since the hydraulic-conductivity estimate is a function of the square of the effective casing radius for all common analysis techniques, it is very important that the appropriateness of the effective casing radius estimate be assessed at each well. A comparison of the H_0^* and H_0 values is the most reliable method for this assessment. If the nominal casing radius (r_{nc}) does not appear appropriate, Stanford et al. (1996) suggest that the effective casing radius (r_c) be estimated using a mass balance of the following form:

$$\pi r_c^2 H_0 = \pi r_{nc}^2 H_0^* \tag{3.1}$$

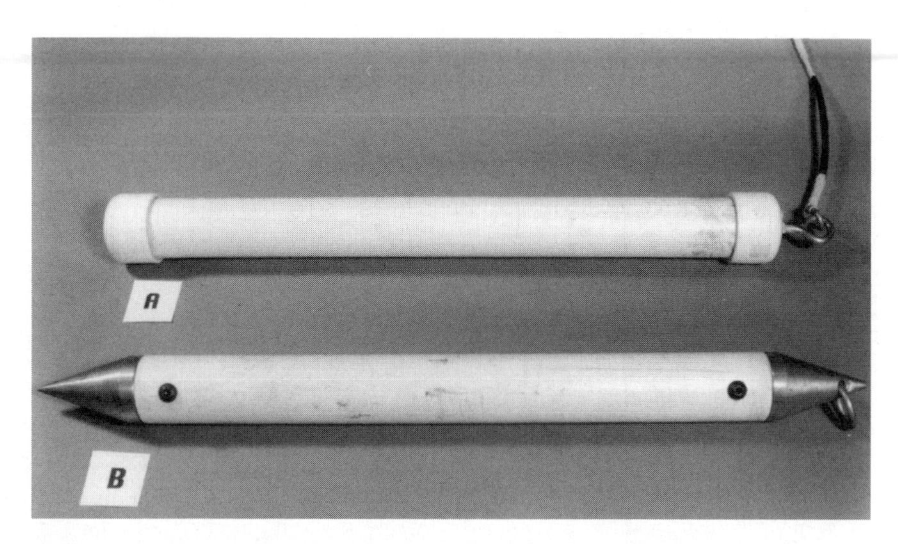

FIGURE 3.3 Conventional (A) and streamlined (B) solid slugs. Note that attempts to make solid slugs more hydrodynamically efficient have met with little success (letter markers are approximately 5 cm wide).

which can be rewritten in terms of r_c as

$$r_c = r_{nc}\sqrt{\frac{H_0^*}{H_0}} \tag{3.2}$$

Modification of this mass balance to incorporate the porosity of the filter pack for the case of wells screened across the water table is discussed in Chapter 6.

In wells in high-permeability formations, Butler et al. (1996) have shown that a comparison of the measured and expected values of the initial displacement can also be used to assess the relative speed of test initiation. Thus, some care must be used in interpreting the difference between H_0 and H_0^* in highly permeable media, as will be discussed further as part of the presentation on the pneumatic method for test initiation.

A reasonable estimate of the initial displacement is also needed for comparisons of normalized data from repeat slug tests performed with different H_0. As discussed in Chapter 2, normalized data plots from repeat tests are a useful tool for the verification of the validity of conventional theory at a particular well. The influence of test initiation should be similar between tests if this approach is to be effective. In the remainder of this section, the primary methods for initiation of a slug test, along with their advantages and disadvantages, will be described.

The most common method for initiating a slug test is to rapidly introduce/remove a solid object (termed a "slug") to/from the well. The solid object is usually a piece of stainless steel or PVC pipe, filled with sand or similar material, and capped at both ends (Figure 3.3A). A hook on the cap at the upper end allows the slug to be

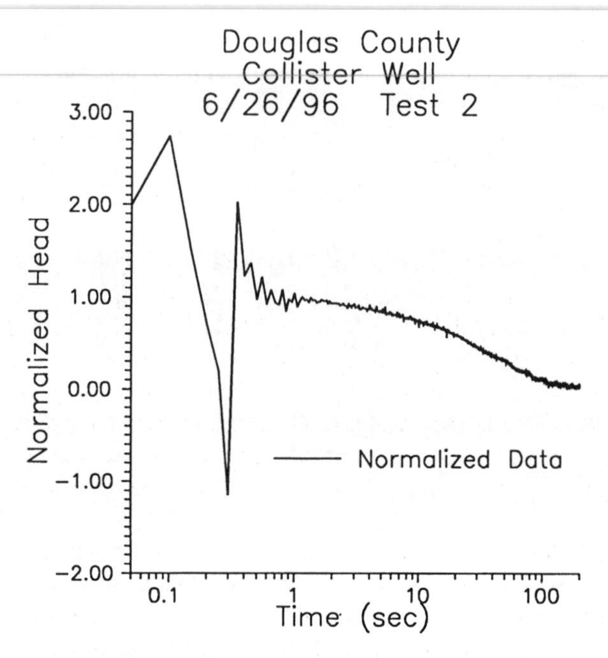

FIGURE 3.4 Normalized head ($H(t)/H_0$, where H(t) is measured deviation from static and H_0 is magnitude of the initial displacement) vs. logarithm of time plot of a rising-head slug test performed with the slug shown in Figure 3.3A. Note that late-time noise was introduced by data acquisition equipment.

attached to a rope or cable. The advantages of initiating a test with a solid slug are that water is not handled during the test, the cost of constructing a slug is very low, a very accurate estimate of H_0^* can be obtained by calibration in blank casing, and a pair of falling- and rising-head tests can be readily performed by introducing and then removing the slug from the water column. There are, however, several disadvantages of this method. First, there may be a considerable amount of early-time "noise" in the response data. Figure 3.4 displays a data record from a test in which a solid slug was removed from the well and a pressure transducer was used to monitor head changes. The large fluctuations in the initial readings from the transducer are a result of at least two phenomena: (1) the movement of the slug causes short-term dynamic pressure disturbances that may be quite large in magnitude; and (2) the transducer and its cable may be entangled with/hit by the slug during the initiation phase. Although the ends of the slug can be streamlined (Figure 3.3B) and, if the well is of large enough diameter, the transducer and its cable can be placed in a smaller diameter pipe (analogous to a stilling well), it is not possible to completely remove this early-time noise. The significance of this noise, however, will depend on the relative speed of the formation response. In the case of Figure 3.4, the formation response is slow enough that the early-time noise has a very limited impact on the test as a whole. However, in extremely permeable formations, this period of early-time noise may last the entire test. Thus, there clearly is an upper limit on the hydraulic conductivity of a formation in which a solid slug can be employed to initiate a slug test. This limit will depend on well geometry, the length of the slug, the speed with which an individual can introduce/remove the slug from the water column, etc.

There are several additional disadvantages to use of a solid slug. For example, the slug may have to be quite long to displace enough water to have a relatively noise-free data record. As its length increases, the slug becomes more difficult to handle and to introduce in a near-instantaneous fashion. Similarly, a solid slug may not be usable in wells with a very short column of water. In this case, the diameter of the slug may have to be relatively large to produce a significant change in water levels. As the diameter of the slug relative to that of the well increases, the chances of interference between the slug and the transducer increases. Moreover, Prosser (1981) describes a series of tests in which the diameter of the slug was so large that it interfered with flow in the casing, producing a hydraulic-conductivity estimate that was over a factor of three less than the best estimate obtained using other initiation methods at the same well. A final disadvantage of a solid slug is that it can serve as a vehicle for cross-hole contamination. At sites of suspected groundwater contamination, the solid slug and its cable, as well as the transducer and its cable, must be appropriately cleaned prior to use in another well.

A second frequently used approach for initiating a slug test is to rapidly introduce/remove water to/from the well. Although the rapid removal of a bailer (a pipe open at the top and with a check valve at the bottom) from the water column is the most common method in this category, water can also be directly poured into the well or can be rapidly pumped out. The two major advantages of this initiation method are that no specialized equipment is required and that an accurate estimate of H_0^* can be obtained by measuring the volume of water added to/removed from the well. There are three significant disadvantages to this approach. First, water must actually be handled with this technique, a critical disadvantage at sites of potential groundwater contamination, where the removed water must be properly disposed of and added water might impact future chemistry measurements. Second, it may be quite difficult to introduce the slug in a near-instantaneous fashion. Both Black (1978) and Prosser (1981) describe slug tests, initiated by pouring water down the well, in which the response data were impacted by the late arrival of water trickling down the casing. Finally, as with the solid slug, test initiation may give rise to considerable dynamic-pressure effects. These effects can be particularly severe when the static water level is relatively deep and the test is initiated by rapidly pouring a large amount of water down the well.

Over the last decade, the pneumatic approach for test initiation has become increasingly popular. This method, which apparently was first fully described in the groundwater literature by Prosser (1981), involves pressurizing the air column in a sealed well by the injection of compressed air or nitrogen gas. This pressurization produces a depression of the water level as water is driven out of the well in response to the increased pressure in the air column (i.e., gas pressure is substituted for water pressure). The water level continues to drop until the magnitude of the total decrease in the pressure head of the water is equal to the magnitude of the total increase in the pressure head of the air column. At that point, the well has returned to equilibrium conditions (i.e., a transducer in the water column has the same reading as prior to

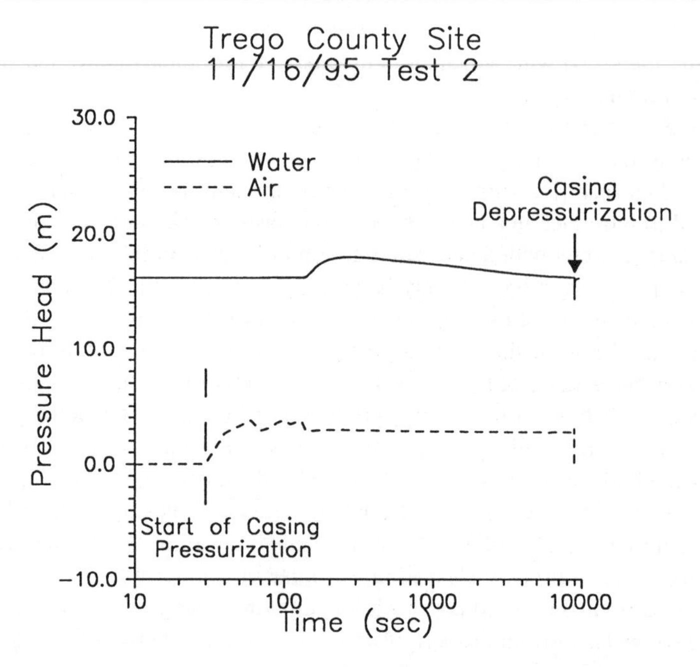

FIGURE 3.5 Pressure head measured by a transducer submerged in the water column (labelled Water) and a transducer at the top of the casing (labelled Air). Note that the lag between the start of casing pressurization and response in the water column is a product of the position of the air-pressure transducer in the well-head apparatus and the depth to water (approximately 132 m).

pressurization) and the test can be initiated by a very rapid depressurization of the air column. Example records of the pressure data recorded by a transducer submerged in the water column and one measuring the pressure in the overlying air column are given in Figure 3.5. As shown in this example, the initial pressurization of the air column will induce an increase in the pressure reading from the submerged transducer because the gas pressure is increased faster than water can flow out of the well. Eventually, however, the pressure reading from the submerged transducer will return to pretest conditions and the test can be initiated. A variant of this approach involves applying a vacuum to the air column and then breaking the vacuum to begin the test (e.g., Orient et al., 1987). The advantages of the pneumatic slug test are that the approach does not require water to be handled during a test, test initiation can be done very rapidly, only the transducer and its cable have to be cleaned before moving to a new well, and a pair of falling- and rising-head tests can be readily performed if both the pressurization (falling-head) and depressurization (rising-head) are done very rapidly with respect to the formation response.

The major disadvantage of the pneumatic approach is that a special air-tight well-head apparatus is required (Figure 3.6). This apparatus has three critical components. First, there must be an air-tight seal around the transducer cable and between the apparatus and the casing. Pressurization of the air column requires that the

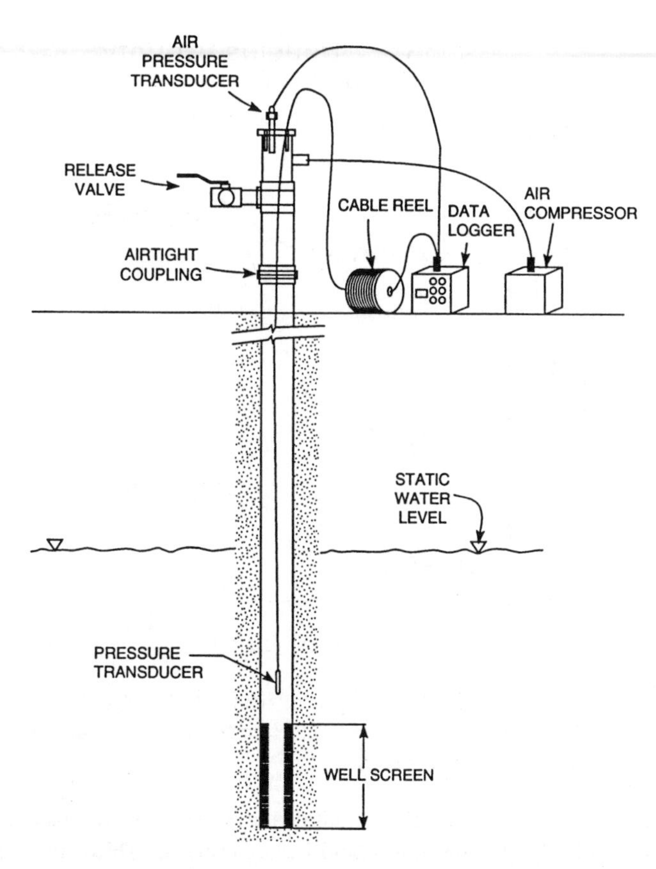

FIGURE 3.6 Hypothetical cross section displaying a well at which the pneumatic method is being used for test initiation. Note that there must be an airtight seal at the point at which the cable for the submerged transducer passes through the top of the well head (figure not to scale). (After McLane et al., 1990.)

volume of gas leaking from the system be much smaller than the volume supplied to the well. Although this is usually not a concern when an air compressor is the source of the injected gas, it can be more of a problem when a compressed-gas canister is employed. Since there may often be a small leak somewhere in the well or at one of the seals, use of a regulator is strongly recommended to maintain a constant air pressure in the casing. Second, the diameter of the valve and piping through which the depressurization occurs should be equal to or larger than the diameter of the well casing. If not, the time required to depressurize the casing may be relatively long. This effect, however, is only a concern in media of very high hydraulic conductivity. As described by Butler et al. (1996), a comparison between the value for the initial displacement measured immediately after depressurization with the submersible pressure transducer (H_0) and the expected value (H_0^*) is an easy way to assess if test initiation can be considered instantaneous relative to the formation response. Figure 3.7 provides two examples of tests performed in media of very high hydraulic conductivity where test initiation cannot be considered instantaneous relative

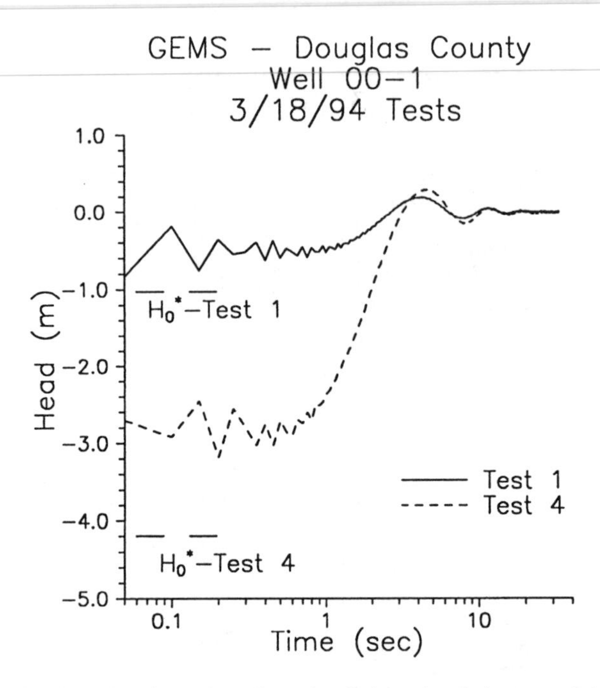

FIGURE 3.7 Head vs. log time plot of a pair of rising-head slug tests initiated with the pneumatic method. In this case, the initiation cannot be considered instantaneous relative to the formation response (H_0^* defined in text).

to the formation response, while Figure 3.8 displays examples of tests in which the instantaneous initiation assumption would be appropriate. Third, there must be a means of measuring H_0^* prior to depressurization. Some authors have recommended the use of one or more downhole electric tapes (e.g., Orient et al., 1987; McLane et al., 1990) to track the position of the water level during the initial pressurization phase, a procedure that may only provide a very rough estimate of the expected initial displacement. An alternate approach is to attach a pressure transducer to the well-head apparatus (e.g., Figure 3.6) and measure the total change in gas pressure from before pressurization until equilibrium conditions are re-established in the well prior to initiation. The expected value of the initial displacement is equal to that total change in pressure.

A second disadvantage of the pneumatic method is that some care must be used when pressurizing the air column. If the water level in the well falls below the top of the well screen, gas can inadvertently be injected into the formation, which can lead to a hydraulic-conductivity estimate that is much lower than the actual value (e.g., Levy et al. 1993). In addition, if the air-water interface moves below the transducer that was initially submerged in the water column, early-time response data will be lost. Fortunately, use of a regulated gas supply and consideration of well geometry and transducer placement can largely prevent such accidents from occurring. Although the need for a supply of compressed gas could be considered a disadvantage, an inexpensive air compressor or a small compressed-gas cylinder, both of which are easily transportable, can readily provide the volume of gas needed

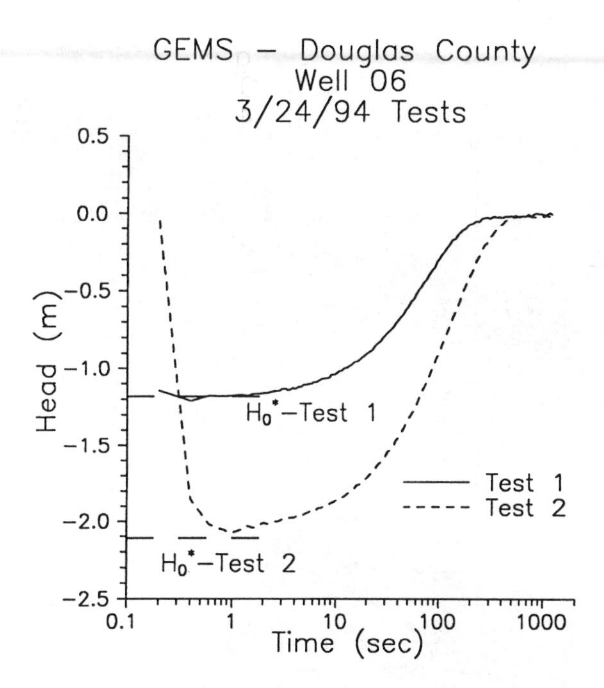

FIGURE 3.8 Head vs. log time plot of a pair of rising-head slug tests initiated with the pneumatic method. In this case, the initiation can be considered instantaneous relative to the formation response (H_0^* defined in text).

for most tests. A final disadvantage of the pneumatic approach is that the time required to re-establish equilibrium conditions prior to depressurization may be quite long in media of low hydraulic conductivity (e.g., Figure 3.5). In order to avoid an often lengthy wait, Shapiro and Greene (1995) propose that the casing be depressurized prior to achieving equilibrium conditions, and present a type-curve approach for the analysis of the resulting response data. Although their method can certainly be used, an alternative approach is to consider water-level responses to the initial pressurization phase as a falling-head slug test, so that both a falling- and rising-head test can be performed at the same well. As discussed in Chapter 2, comparison of responses from falling- and rising-head tests at the same well can provide very useful information about the viability of the conventional set of assumptions invoked for the analysis of slug-test response data.

A final relatively common method for test initiation is the use of a downhole packer-based system (Figure 3.9). In this approach, a packer is lowered beneath the static level in the well. The central pipe upon which the packer is mounted is closed and/or the packer is inflated to isolate the portion of the well above the packer from that below. Water can then be removed from/added to the casing above the packer, or the water level above the packer can be raised/lowered with a solid slug. A test is initiated either by deflating the packer (e.g., Patterson and Devlin, 1985; Priddle, 1989) or by opening the central pipe (e.g., McElwee and Butler, 1989). Packer deflation may take a few to several seconds, thus restricting the deflation method to

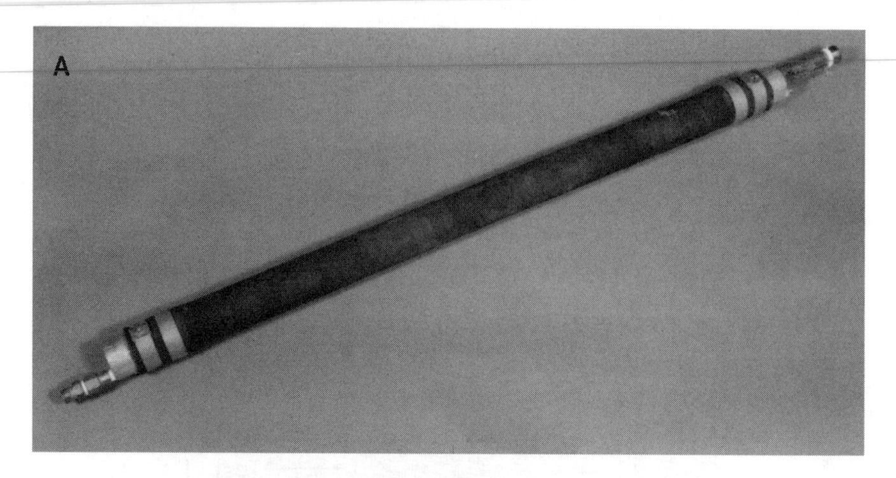

FIGURE 3.9 Inflatable (A) and mechanical (B) packers. Note that the inflatable packer is of the fixed-head type (outer diameters of inflatable and mechanical packers are approximately 5 cm and 10 cm, respectively).

media of moderate to low hydraulic conductivity. Opening the central pipe upon which the packer is mounted, however, can be done extremely quickly either electronically via a solenoid valve or mechanically with a wireline or small-diameter steel rods. The major advantages of packer-based methods are that a test can be initiated quite rapidly, a very good estimate of H_0^* can be obtained, and head recovery below the packer is relatively fast when the central pipe is closed. This relative rapid recovery of head below a closed packer, which is a result of the recovery being a function of the compressibility of water and test equipment and not the radius of the well casing (Neuzil, 1982), greatly diminishes the impact of pressure disturbances caused by the lowering/inflation of the packer or from a preceding slug test.

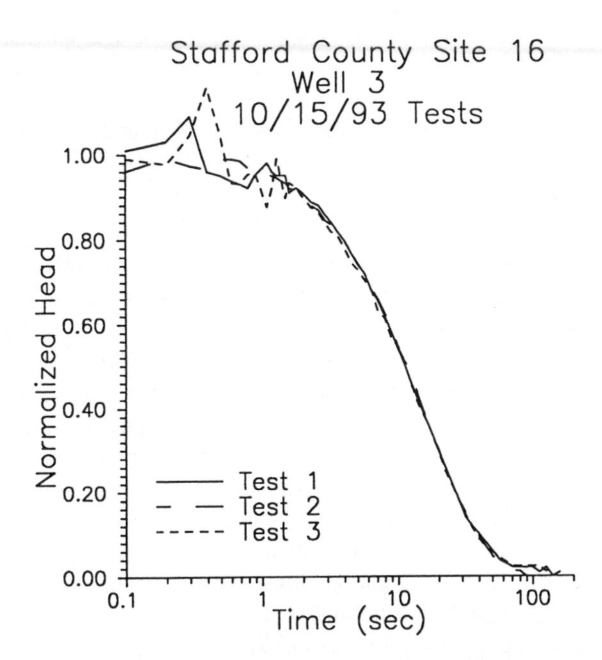

FIGURE 3.10 Normalized head ($H(t)/H_0$) vs. log time plot of a series of slug tests initiated by opening the central pipe of a packer.

Thus, repeat tests can be performed at a greater frequency than with other initiation methods. Although a slug test can be initiated quite rapidly by opening the central pipe of the packer, there will usually be a short interval, approximately 1 to 2 seconds in duration, in which dynamic-pressure effects produced by test initiation are affecting transducer measurements. Figure 3.10 shows data from a series of slug tests performed in a well in a sand and gravel aquifer that display dynamic-pressure effects that are typical of what is seen in the field. This early-time noise is essentially only a problem in small-diameter wells sited in media of extremely high hydraulic conductivity, where the duration of the dynamic pressure effects may be a significant fraction of the total length of the test. In that situation, however, none of the techniques described here are capable of initiating a test in a manner that can be considered near-instantaneous relative to the formation response.

The major disadvantages of packer-based methods are that packers can be expensive, the central pipe upon which the packer is mounted may restrict flow, and the packer must be cleaned prior to use at another well. Although inflatable packers (Figure 3.9A) can be expensive, mechanical packers (Figure 3.9B) are a very inexpensive alternative for situations where the packer is to be positioned in the cased portion of shallow wells. If inflatable packers are necessary, such as in a straddle-packer system for multilevel slug tests (e.g., Butler et al., 1994), sliding-head packers are best because the same packer can be used in wells of a range of diameters. A significant advantage of a mechanical packer is that the central pipe is usually larger than that in an inflatable packer designed for the same size well. This, however, is only important in media of high hydraulic conductivity where the central pipe of the packer may restrict flow. A comparison of normalized data from repeat tests performed at the same well using different H_0 (difference on the order of a factor of two or greater) will reveal if the central pipe is restricting flow. If these normalized data approximately coincide, then the pipe diameter is not affecting test responses (see Chapter 8).

DEGREE OF HEAD RECOVERY

The necessity of performing repeat slug tests at a well to verify the validity of conventional theory is one of the themes of this book. A natural question to ask therefore is what degree of head recovery is required prior to repeating a test. Since a number of the mathematical models on which the analysis methods are based assume that formation heads are at equilibrium conditions prior to test initiation, a conservative approach would be to wait until static conditions have been reestablished before repeating a test. However, given that the process of head recovery during a slug test can be approximately represented as an exponential decay, the time required for recovery of the last few percent of the original deviation from static can be quite long. Although this does not present much of a problem in units of high hydraulic conductivity, it can greatly lengthen the duration of a program of slug tests in less permeable formations. Thus, an issue of considerable practical importance is that of how much error is introduced into the hydraulic-conductivity estimate when a test is repeated prior to complete recovery.

Apparently, this issue of the degree of head recovery has not been fully assessed in a field setting. Shapiro and Greene (1995) describe a method for the analysis of pneumatic slug tests initiated before complete recovery from the pressurization of the air column. Their focus, however, is on the development of a new set of type curves; so, they do not explicitly address the issue of the error introduced by analysis with conventional methods when a test is started prior to complete recovery. Butler (1997) presents the results of a theoretical analysis of the effect of incomplete recovery based on the principle of superposition. Figure 3.11 displays results from that analysis for the case of a well that is fully screened across the formation. These results, which are appropriate for hydraulic-conductivity estimates obtained with the methods of Hvorslev (1951) and Bouwer and Rice (1976), indicate that the effect of incomplete recovery is a function of α, the dimensionless storage parameter. Figure 3.11 shows that, for moderate to small values of α, agreement between parameter estimates from repeat tests will be within 10% when the deviation from static at the time of test initiation (henceforth designated as the residual deviation) is less than 20% of H_0. For large α values, the residual deviation should be less than 5% of H_0 for parameter estimates from repeat tests to agree within 10%. It is important to emphasize two aspects of these results. First, these results were obtained assuming that repeat slug tests were performed using the same H_0. If H_0 changes between tests, the residual deviation should be evaluated relative to the H_0 for the second test. Second, these results were obtained for a hypothetical well assumed to be fully screened across the formation. As Butler (1997) demonstrates, the effect of incomplete recovery is diminished in a well that is only screened across a portion of the formation. Thus, in virtually all cases of practical significance, very little error

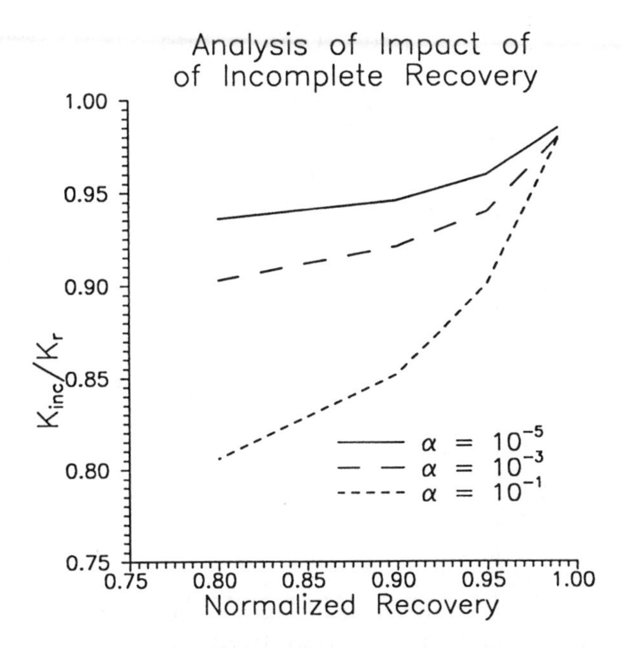

FIGURE 3.11 Plot of hydraulic conductivity ratio (conductivity estimated for conditions of incomplete recovery (K_{inc}) over actual conductivity (K_r)) vs. normalized recovery (H_{inc}/H_0, where H_{inc} is the residual deviation) as a function of α (α defined in text).

will be introduced into parameter estimates by performing repeat tests when the residual deviation is less than 5% of H_0. If there is concern that a test may have been impacted by incomplete recovery, the response data from that test should be plotted as the log of the deviation of head from static conditions vs. the time since test initiation. If that plot is near linear in form, such as that shown in Figure 3.12, the effects of incomplete recovery should be quite small.

Two additional points need to be made about the issue of incomplete recovery. First, although the analyses described in the previous paragraph indicate that repeat tests can be performed before complete recovery, head measurements should always be taken prior to and at the end of a program of slug tests at a particular well to assess if tests are being impacted by temporal trends in the assumed stationary background head. Second, a useful characteristic of a packer system for test initiation is that the central pipe of the packer can be closed to terminate the test prior to complete recovery. Since head recovery below the closed packer will be a function of the compressibility of water and the packer, recovery will be much more rapid than if the central pipe had remained open. A pressure sensor can be located below the packer to assess when sufficient recovery has occurred.

PERFORMANCE GUIDELINES

The following four performance guidelines can be extracted from the discussions of this chapter:

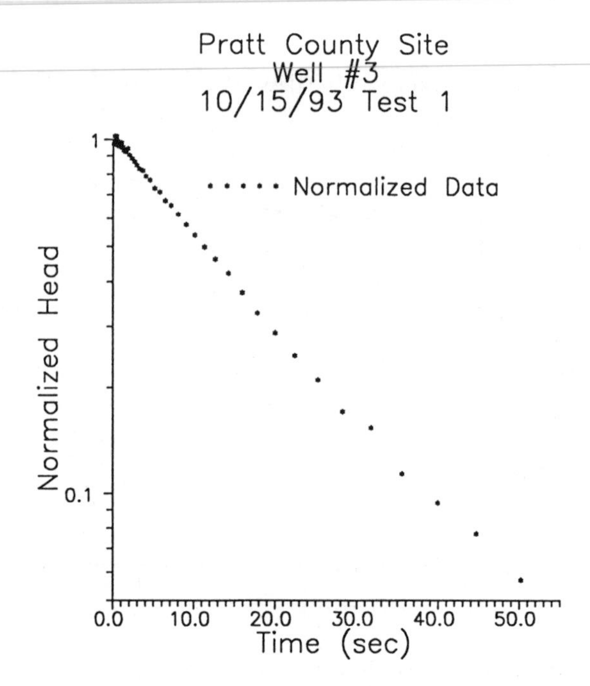

FIGURE 3.12 A near-linear logarithm of normalized head ($H(t)/H_0$) vs. time plot.

1. The equipment used to obtain and store head measurements during the course of a slug test must be appropriate for the expected head response. In formations of moderate to high hydraulic conductivity, a pressure transducer connected to a data-acquisition device is the best option. In less permeable formations, use of an electric tape is also acceptable. Transducer operation should always be checked prior to and after a program of tests at a particular well;
2. A slug test should be initiated in a manner that can be considered near instantaneous relative to the formation response. A near-instantaneous initiation is needed to satisfy assumptions invoked in some methods for the analysis of slug-test response data. In formations of very high hydraulic conductivity, test initiation using pneumatic or packer-based systems is most appropriate. Test initiation using a solid slug is a reasonable alternative in less permeable media. Techniques that involve the addition or removal of water should be avoided;
3. Accurate estimates of both the expected (H_0^*) and measured (H_0) values for the initial displacement are required. A comparison between these two quantities can be used to assess the appropriateness of the effective casing radius estimate and for a check on the relative speed of test initiation. If the nominal casing radius does not appear to be a reasonable estimate of the effective casing radius, a mass balance can be used to calculate a more appropriate value. In addition, an accurate H_0 estimate allows ready comparison between

repeat slug tests performed at a well, enabling the viability of conventional theory to be assessed for tests at that well;

4. The residual deviation from static should be less than 5% of H_0 before a slug test is repeated at a well. If H_0 is changed between tests, the recovery criterion should be applied using the H_0 for the second test. If there is uncertainty about whether incomplete recovery is affecting test responses at a particular well, the test data should be plotted as the log of the deviation of head from static conditions vs. time. A near-linear form for this plot indicates that the residual deviation is not significantly affecting test responses at that well.

4 Pre-Analysis Processing

CHAPTER OVERVIEW

At the end of a slug test, one has a record of the change of head in the well in response to the slug-induced disturbance. However, before these data can be used to estimate the hydraulic properties of the formation, they must undergo additional processing. The focus of this chapter will be on the major elements of this pre-analysis processing. Particular emphasis will be placed on the use of data obtained with a pressure transducer. The chapter will conclude with the definition of a series of practical guidelines for the pre-analysis processing of head data collected during a slug test.

PRE-ANALYSIS DATA PROCESSING

The two major activities in the pre-analysis processing phase are the conversion of sensor measurements into normalized response data and the estimation of the magnitude of the initial displacement (H_0) and the time of test initiation (t_0). These and related issues are discussed in the following two sections.

DATA CONVERSION

The response data from a slug test usually consist of depth-to-water measurements obtained with an electric tape or, more commonly, measurements of the pressure exerted by the overlying column of water obtained using a pressure transducer. Because the pressure transducer is the most frequently used device for acquiring response data, the focus of this discussion will be on the processing of transducer data. The extension to measurements obtained with an electric tape or other devices is quite straightforward.

The particular form of the transducer data will depend on the nature of the data acquisition device to which the transducer is connected. The data can be in one of three possible forms: (1) the unprocessed output from a transducer (digitized current or voltage measurements), (2) the pressure head produced by the column of water overlying the transducer, or (3) the deviation of the total head from static conditions. Prior to analysis, all measurements must be converted into this third form, the deviation of head from static conditions. The primary steps in this conversion process are briefly described here.

When the response data consist of the unprocessed output from a transducer, the first step is to convert these voltage or current measurements into physical units of pressure head. This conversion is normally done using the following linear equation:

$$hp_i = mx_i - b \tag{4.1}$$

where: hp_i = pressure head produced by the overlying column of water, [L];
x_i = current or voltage measurement from transducer, [I] or [V];
m = slope of conversion equation, [L/I] or [L/V];
b = intercept of conversion equation, [L].

The values of the slope and intercept parameters (henceforth designated as the calibration parameters) are determined by periodic calibration under carefully controlled laboratory conditions, and should be checked in the field as described in Chapter 3.

Once the transducer output has been converted into units of pressure head, the next step is to calculate the deviation of head from static (background) conditions. This calculation is done in a two-stage process. The first stage consists of an examination of the test data to determine what pressure head value corresponds to static conditions, while the second stage is the subtraction of this value from all pressure head measurements:

$$\begin{aligned} hp_i - hp_{st} = h_i &= m(x_i - x_{st}) - (b - b) \\ &= m(x_i - x_{st}) \end{aligned} \tag{4.2}$$

where: hp_{st} = pressure head at static conditions, [L];
x_{st} = current or voltage measurement for static conditions, [I] or [V];
h_i = deviation of total head from static conditions, [L].

There are two noteworthy ramifications of the subtraction operation performed in Equation (4.2). First, the deviation of the pressure head from static conditions is equal to the deviation of the total head from static because the elevation head component, which is the position of the transducer, should not change during a test. Second, as shown in Equation (4.2), the intercept parameter is eliminated by the subtraction operation. Thus, errors in the intercept parameter will have no influence on the calculated deviations from static as long as the intercept parameter is not a function of pressure.

The next step in the pre-analysis processing is to normalize the deviation data by the initial displacement from static (H_0). Prior to normalization, an estimate of H_0 must be obtained from an examination of the data record. Once H_0 has been estimated, the deviation data are divided (normalized) by that value:

$$\frac{h_i}{H_0} = \frac{m(x_i - x_{st})}{m(x_0 - x_{st})} = \frac{x_i - x_{st}}{x_0 - x_{st}} \tag{4.3}$$

where: x_0 = current or voltage measurement for pressure head equivalent to H_0, [I] or [V].

There is a very significant ramification of the division operation performed in Equation (4.3). As shown in the equation, this operation removes the influence of the slope parameter from the normalized data. Thus, the normalized data are independent of the specific values assigned to either of the calibration parameters. This independence from the calibration parameters is a significant advantage of working with normalized data, as uncertainty concerning these parameters can often introduce error into the response data. It is important to emphasize, however, that the underlying assumption of Equation (4.3) is that the calibration parameters are constant over the range of pressure heads observed during a test. The viability of this assumption can be readily checked in the field prior to and/or at the end of a program of tests. This is done by systematically changing the position of the transducer in the water column so that the transducer response can be assessed over a range of pressure heads approximating that observed during the slug tests. At least three pressure head measurements should be taken at approximately equal increments over this range, after which calibration parameters can be calculated for all possible pairs of measurements. Close agreement between the parameters calculated in this fashion is a demonstration of the validity of the assumption that the calibration parameters are independent of pressure.

The final step in the pre-analysis processing is to reinitialize the time record to the actual time at which the test began (t_0). In many slug tests, data collection begins prior to, and not concurrent with, the actual start of the test (e.g., Figure 2.7A). Thus, the times corresponding to the pressure measurements must be adjusted so that they are a measure of actual test duration. As in the case of H_0, an estimate of t_0 can be found from an examination of the data record. Once an estimate for t_0 has been obtained, that value is subtracted from every time entry in the data record. The reinitialized data record is then truncated by eliminating all negative time entries.

The final product of the pre-analysis processing is a record of the deviation of head from static conditions that has been normalized by H_0 and reinitialized to t_0. As discussed in detail in Chapter 2, normalized response data from a series of repeat tests can be used to assess the viability of some of the critical assumptions underlying the conventional methods for the analysis of slug-test data. The comparison of normalized data from a series of tests is thus a critical step in the selection of a theoretical model for use in the subsequent analysis. Although the normalized data do not necessarily have to be utilized in the actual analysis process, use of normalized data is highly recommended whenever the response data are collected with a pressure transducer.

Estimation of H_0 and t_0

In the preceding section, it was assumed that the magnitude of the initial displacement and the time of test initiation could be readily estimated from the data record. In tests performed in formations of moderate to low hydraulic conductivity, the estimation of these quantities is a very straightforward process. However, in more

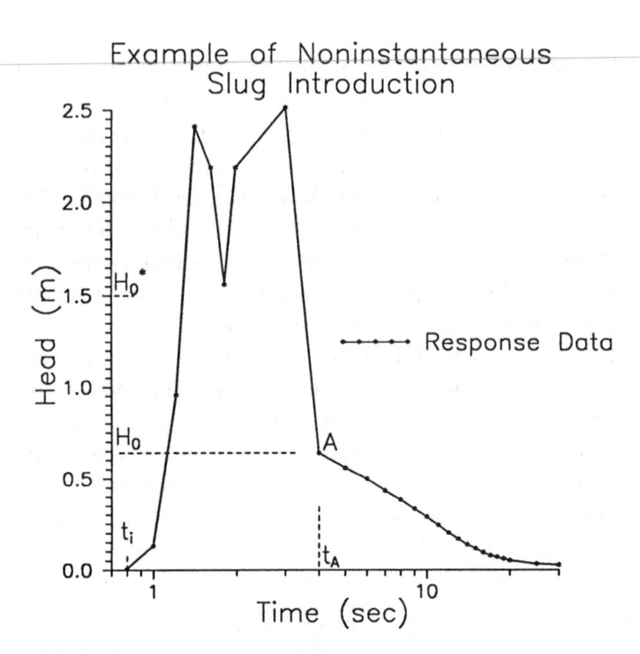

FIGURE 4.1 Head vs. log time plot of a test initiated with a solid slug. Test initiation is not instantaneous relative to the formation response. H_0^*, t_i, t_A, and A are defined in text. (After Butler, 1996.)

permeable formations, these quantities can be rather difficult to estimate as a result of early-time "noise" in the data record, produced by a number of factors related to test initiation, in conjunction with a very rapid formation response (e.g., Figure 4.1). The basic problem for tests such as the one depicted in Figure 4.1 is that the initial displacement is introduced in a manner that is not instantaneous relative to the formation response. There are three common approaches for estimation of H_0 and t_0 from response data that are impacted by noninstantaneous slug introduction. One approach (henceforth designated as the translation method following the terminology of Pandit and Miner [1986]) is to ignore the noisy early-time portions of the record and assume that the test started at point A on Figure 4.1. The head (H_0) and time (t_A) of point A are then used for the initial displacement and start time of the test, respectively. A second approach is to use the expected initial displacement (H_0^*) and the actual time at which the slug introduction began (t_i), but just ignore the early-time portions of the record. A third approach is to use H_0 and t_i, but again ignore the early-time portions of the record.

Butler (1996) assessed the potential of all three approaches with a numerical model. The primary finding of that investigation was that the translation method is clearly superior to the other alternatives. However, the translation method is only appropriate when a plot of the logarithm of the response data vs. time is approximately linear. In cases where the response plot has a pronounced concave-upward curvature, the translation method can introduce considerable error into the hydraulic-conductivity estimate. In those conditions, every effort should be made to ensure that the slug is introduced in a near-instantaneous fashion, or the noninstantaneous slug introduction should be explicitly modelled as part of the analysis procedure. Similar issues related to the processing of oscillating responses are discussed in Chapter 8. Note that this evaluation of the translation method used the Hvorslev method (see Chapter 5) to obtain estimates of the hydraulic conductivity of the formation. Since, as will be discussed in Chapter 5, the Hvorslev method is based on a quasi-steady-state representation of the slug-induced flow, the translation method involves no additional simplifications of the mathematical model of the flow system. However, for more rigorous analysis methods, such as those based on the Cooper et al. and KGS models (see Chapters 5 and 6), the translation method is not compatible with the underlying mathematical model. These more rigorous methods are often employed in screening analyses to detect the presence of a well skin or to assess the relative significance of vertical flow (see Chapters 5, 9, and 12). Use of the translation method can hinder the effectiveness of these techniques as screening tools. Thus, as a general rule, efforts should be made to initiate a slug test as rapidly as possible to avoid use of the translation method except in the case of highly permeable formations (see Chapter 8).

PRE-ANALYSIS PROCESSING GUIDELINES

The following three guidelines for the pre-analysis processing of response data can be extracted from the material of this chapter:

1. The response data should be transformed into the normalized deviation from static. Normalized data from a series of repeat tests can be used to assess the appropriateness of conventional theory as described in Chapter 2;
2. The normalized deviation data should be utilized in the analysis if a transducer was used to collect the response data. If this is not done, errors in the calibration parameters can introduce error into the parameter values estimated from the response data;
3. The translation method should be used in cases of noninstantaneous slug introduction. The appropriateness of this approach for any particular test can be assessed using a plot of the logarithm of the response data vs. time. However, the translation method can limit the effectiveness of screening analyses, so a considerable effort should be made to initiate a test as rapidly as possible.

5 The Analysis of Slug Tests — Confined Formations

CHAPTER OVERVIEW

Upon completion of the preliminary processing, the response data are ready to undergo formal analysis. The objective of the analysis phase is to obtain estimates of the hydraulic conductivity and, under certain conditions, the specific storage of the portion of the formation opposite the screened or open section of the well. The emphasis of this chapter will be on the major techniques for the analysis of response data from slug tests performed in confined formations. Specialized techniques developed for tests in confined formations in which the hydraulic conductivity is very high or very low, or in cases where the response data are impacted by a well skin, are discussed in later chapters.

THEORETICAL MODELS FOR THE ANALYSIS OF RESPONSE DATA

In this and following chapters, the major methods for the analysis of response data are described. Each of these methods is based on a series of assumptions regarding the nature of the slug-induced flow. A particular set of assumptions invoked for a given well-formation configuration constitutes the model that the analyst has adopted for that specific hydrogeologic setting. The analysis methods are grouped here on the basis of the fundamental assumptions incorporated in the theoretical models underlying the various techniques. In cases where a particular method could be discussed in several chapters, the method is described in the chapter deemed to be most closely associated with the fundamental characteristics of that technique. Thus, for example, methods that have been specifically developed for analysis of slug tests in highly permeable aquifers will be discussed in the chapter on tests in high-conductivity formations, rather than in chapters on tests in confined or unconfined formations. Many of the analysis methods were originally developed as manual curve fitting procedures. Although manual approaches are utilized less frequently in this era of automated analyses, the description of the analysis techniques given here will emphasize the graphical procedures on which they are based. The rationale for this approach is that the better the understanding of the details of a particular method, the better the interpretation of the results produced by that technique. Note that practical guidelines for the analysis of slug tests will not be given at the end of each chapter. These guidelines instead will be presented in a separate summary chapter (Chapter 12) following the discussion of all analysis methods.

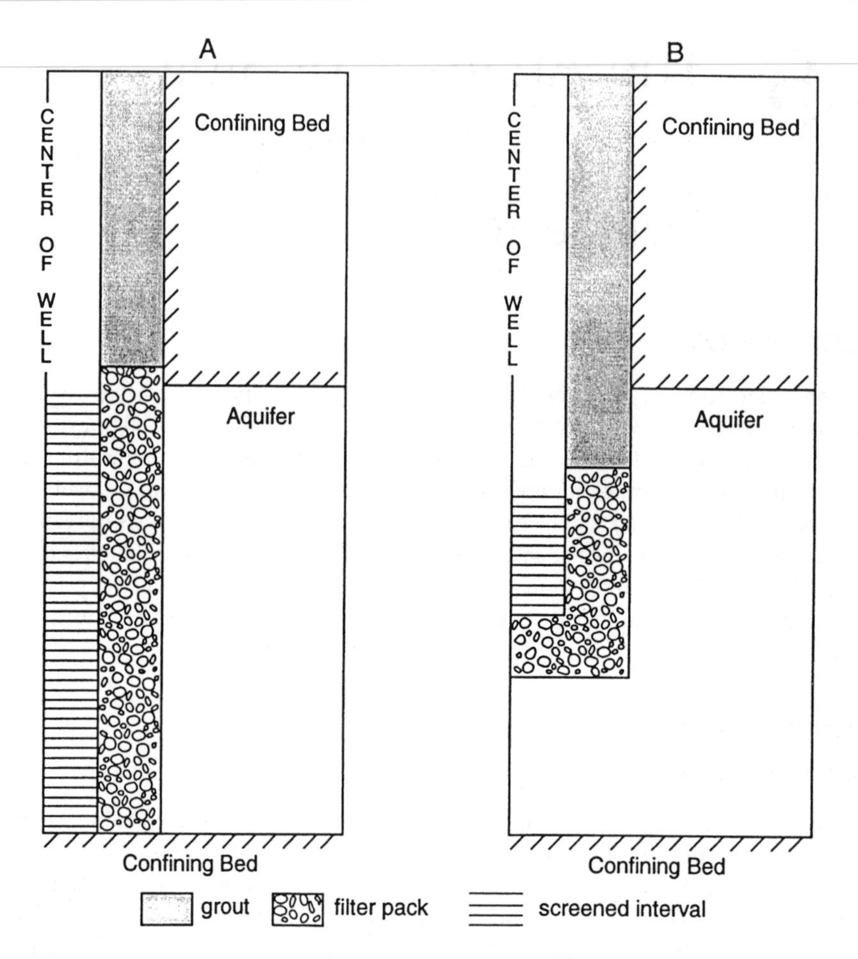

FIGURE 5.1 Hypothetical cross sections depicting a fully penetrating (A) and partially penetrating (B) well in a confined aquifer. (Figures not to scale.)

SLUG TESTS IN CONFINED FORMATIONS

The major methods for the analysis of response data from slug tests in confined formations can all be discussed in the context of the hypothetical cross sections depicted in Figure 5.1. In all cases, the confining beds bounding the formation in the vertical plane are assumed to be impermeable (i.e., the formation is perfectly confined). The methods are primarily classified on the basis of the length of the well screen relative to the thickness of the formation. A well that is screened over the full thickness of the formation (Figure 5.1A) is called a fully penetrating well, while the term partially penetrating is used to designate a well that is screened over a limited portion of the formation (Figure 5.1B). In both cases, the methods can be further subdivided on the basis of their conceptualization of the storage properties of the media, as the elastic storage mechanisms encapsulated in the specific storage parameter are neglected in several methods.

Fully Penetrating Wells

The vast majority of slug tests performed in fully penetrating wells in confined formations are analyzed using the method of Cooper et al. (1967) or that of Hvorslev (1951). A more recent technique, the approximate deconvolution method of Peres et al. (1989), also has considerable potential for analysis of tests performed in this configuration. These three techniques are described in the following sections. After the description of these techniques, other less commonly utilized approaches are briefly summarized.

The Cooper et al. Method

The Cooper et al. method is based on the mathematical model defined as follows:

$$\frac{\partial^2 h}{\partial r^2} + \frac{1}{r}\frac{\partial h}{\partial r} = \frac{S_s}{K_r}\frac{\partial h}{\partial t} \tag{5.1a}$$

$$h(r,0) = 0, r_w < r < \infty \tag{5.1b}$$

$$H(0) = H_0 \tag{5.1c}$$

$$h(\infty,t) = 0, t > 0 \tag{5.1d}$$

$$h(r_w,t) = H(t), t > 0 \tag{5.1e}$$

$$2\pi r_w K_r B \frac{\partial h(r_w,t)}{\partial r} = \pi r_c^2 \frac{dH(t)}{dt}, t > 0 \tag{5.1f}$$

where:
- h = deviation of hydraulic head in the formation from static conditions, [L];
- K_r = radial component of hydraulic conductivity, [L/T];
- S_s = specific storage, [1/L];
- B = formation thickness, [L];
- H = deviation of head in well from static conditions, [L];
- H_0 = magnitude of initial displacement, [L];
- r_w = effective radius of well screen, [L];
- r_c = effective radius of well casing, [L];
- t = time, [T];
- r = radial direction, [L].

The mathematical model defined by Equations (5.1a) to (5.1f) is based on a series of assumptions about the slug-induced flow system. These assumptions range from those of a homogeneous formation and Darcian flow (Equation [5.1a]) to those of instantaneous slug introduction (Equations [5.1b] and [5.1c]) and negligible well losses [5.1e]). Two assumptions of particular significance with respect to a discussion of differences between the Cooper et al. and Hvorslev methods are (1) hydrologic boundaries in the plane of the formation are at a great distance from the test well (Equation [5.1d]), and (2) the elastic storage mechanisms represented by the specific storage parameter affect test responses (Equation [5.1a]).

As shown by Cooper et al. (1967), the analytical solution to the mathematical model defined in Equations (5.1a) to (5.1f) can be written as:

$$\frac{H(t)}{H_0} = f(\beta, \alpha) \tag{5.2}$$

where: $\beta = K_r Bt/r_c^2$, commonly designated as the dimensionless time parameter;
$\alpha = (r_w^2 S_s B)/r_c^2$, commonly designated as the dimensionless storage parameter.

This solution, when plotted as normalized head vs. the logarithm of β, forms a series of type curves, with each type curve corresponding to a different value of α (Figure 5.2). The Cooper et al. method involves fitting one of the α curves to the field data via manual curve matching or an automated analog. The method essentially consists of the following five steps:

1. The normalized response data are plotted vs. the logarithm of the time since the test began as shown in Figure 5.3A;
2. The data plot is overlain by a type-curve plot prepared on graph paper of the same format (i.e., number of log cycles). The type curves are then moved parallel to the x axis of the data plot until one of the α curves approximately matches the plot of the field data as shown in Figure 5.3B. Note that the y axes are not shifted with respect to one another during this process;
3. Match points are selected from each plot. For convenience's sake, β is set to 1.0 and the real time ($t_{1.0}$) corresponding to $\beta = 1.0$ is read from the x axis of the data plot (e.g., Figure 5.3B). An α estimate (α_{cal}) is obtained from the type curve most closely matching the data plot ($\alpha_{cal} = 10^{-5}$ in Figure 5.3B);
4. An estimate for the radial component of hydraulic conductivity is calculated from the definition of β for $\beta = 1.0$:

$$K_r = \frac{r_c^2}{Bt_{1.0}} \tag{5.3}$$

5. An estimate for the specific storage is calculated from the definition of α:

$$S_s = \frac{\alpha_{cal} r_c^2}{r_w^2 B} \tag{5.4}$$

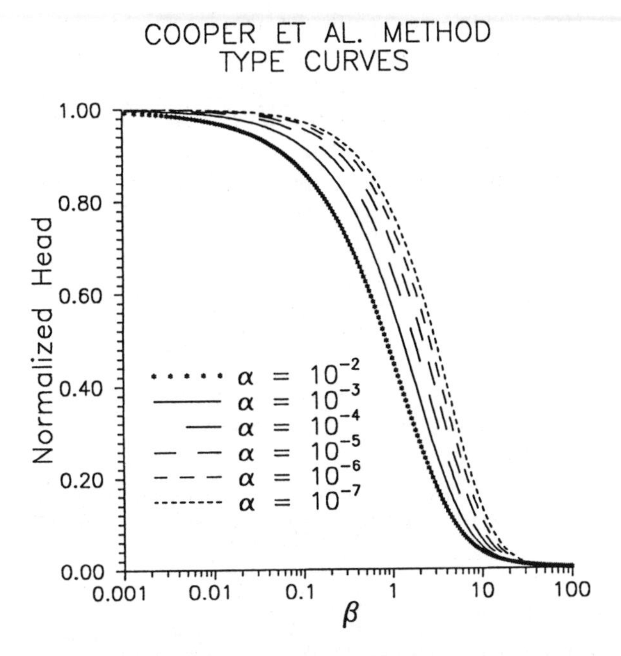

FIGURE 5.2 Normalized head ($H(t)/H_0$, where $H(t)$ is deviation from static and H_0 is magnitude of the initial displacement) vs. the logarithm of β, the dimensionless time, type curves generated with the Cooper et al. model. Each curve labelled using α, the dimensionless storage parameter (α and β defined in text).

where: α_{cal} = α value calculated via curve matching.

An example can be used to illustrate this procedure. In June of 1991, a series of slug tests were performed at a monitoring well in Lincoln County, Kansas (Butler and Liu, 1997). The well was screened in what was interpreted to be a deltaic sequence consisting of mudstone interbedded with very fine sandstone. Table 5.1 summarizes the well-construction information. Note that the effective screen radius was set equal to the radius of the filter pack following the rationale outlined in Chapter 2. Table 5.2 lists the test data employed in the analysis, while Figure 5.4A is a plot of that data in a normalized head vs. log time format. When the curve-matching process was performed, a very close match was found between the data plot and the $\alpha = 0.0125$ type curve as shown in Figure 5.4B. Substituting the $t_{1.0}$ value found from the type curve match into Equation (5.3) yields an estimate of the radial component of hydraulic conductivity equal to 3.48×10^{-4} m/d, a value slightly higher than that reported by Butler et al. (1996) as a result of a reinterpretation of the effective screen length parameter in keeping with the discussion of Chapter 2. Substituting 0.0125 for α_{cal} in Equation (5.4) yields an estimate of specific storage equal to 2.97×10^{-4} m^{-1}, a value that is reasonable for the materials through which the well is screened.

Although the Cooper et al. technique does enable a value to be estimated for the specific storage of the formation, there is often considerable uncertainty regarding

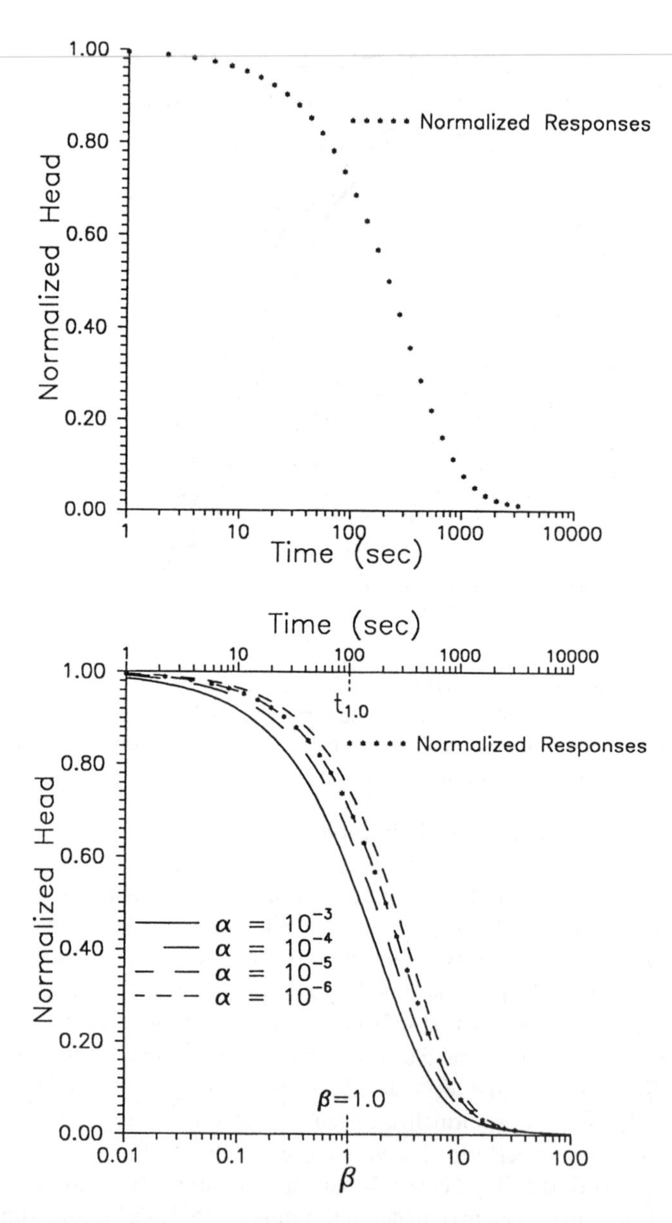

FIGURE 5.3 (A) Normalized head ($H(t)/H_0$) vs. log time plot simulated with the solution of Cooper et al.; (B) Example of the type curve matching procedure employed in the Cooper et al. method. ($t_{1.0}$ and $\beta_{1.0}$ are defined in text.)

that estimate. One reason for this can be readily seen from a close examination of Figure 5.2. Note that as α decreases in magnitude the shape of the type curves becomes quite similar, making it very difficult, on the basis of shape alone, to distinguish between α curves that differ by one to two orders of magnitude. This lack of sensitivity of test responses to α was first noted by Cooper et al. (1967) and

TABLE 5.1
Well Construction Information for Well Ln-1

Well Designation	r_w(m)	r_c(m)	B(m)
Lincoln County Well Ln-1	0.071	0.025	3.05

TABLE 5.2
Response Data from 6/14/91 to 6/18/91 Slug Test in Well Ln-1 at Lincoln County Monitoring Site

Time (s)	Head (m)	Normalized Head
3.0	10.339	0.999
6.0	10.339	0.999
9.2	10.339	0.999
12.3	10.328	0.998
15.6	10.339	0.999
18.9	10.339	0.999
22.3	10.328	0.998
25.8	10.328	0.998
29.4	10.328	0.998
36.8	10.317	0.997
40.7	10.317	0.997
48.6	10.306	0.996
56.9	10.296	0.995
65.6	10.285	0.994
74.8	10.285	0.994
89.3	10.263	0.992
104.9	10.252	0.991
121.6	10.230	0.988
139.5	10.219	0.987
158.7	10.208	0.986
186.4	10.187	0.984
216.9	10.165	0.982
259.2	10.132	0.979
296.8	10.110	0.977
348.9	10.078	0.974
407.5	10.056	0.972
473.3	10.023	0.968
547.3	9.990	0.965
648.1	9.947	0.961
743.6	9.914	0.958

TABLE 5.2 (continued)
Response Data from 6/14/91 to 6/18/91 Slug Test in Well Ln-1 at Lincoln County Monitoring Site

Time (s)	Head (m)	Normalized Head
873.8	9.871	0.954
1023.6	9.827	0.949
1195.8	9.773	0.944
1393.8	9.718	0.939
1621.4	9.653	0.933
2277.9	9.459	0.914
2517.9	9.401	0.908
2997.9	9.287	0.897
3477.9	9.177	0.887
4197.9	9.023	0.872
4677.9	8.933	0.863
5637.9	8.763	0.847
6357.9	8.634	0.834
7557.9	8.430	0.815
8757.9	8.233	0.795
10198.0	8.001	0.773
11878.0	7.746	0.748
13798.0	7.440	0.719
15958.0	7.126	0.689
18598.0	6.780	0.655
21718.0	6.453	0.623
25318.0	6.128	0.592
29398.0	5.798	0.560
34198.0	5.457	0.527
39958.0	5.085	0.491
46438.0	4.713	0.455
54118.0	4.330	0.418
63238.0	3.935	0.380
73798.0	3.534	0.341
85798.0	3.134	0.303
100200.0	2.724	0.263
116760.0	2.351	0.227
135960.0	2.009	0.194
158520.0	1.668	0.161
184920.0	1.348	0.130
215640.0	1.050	0.101
251400.0	0.813	0.079
292920.0	0.620	0.060
341640.0	0.468	0.045

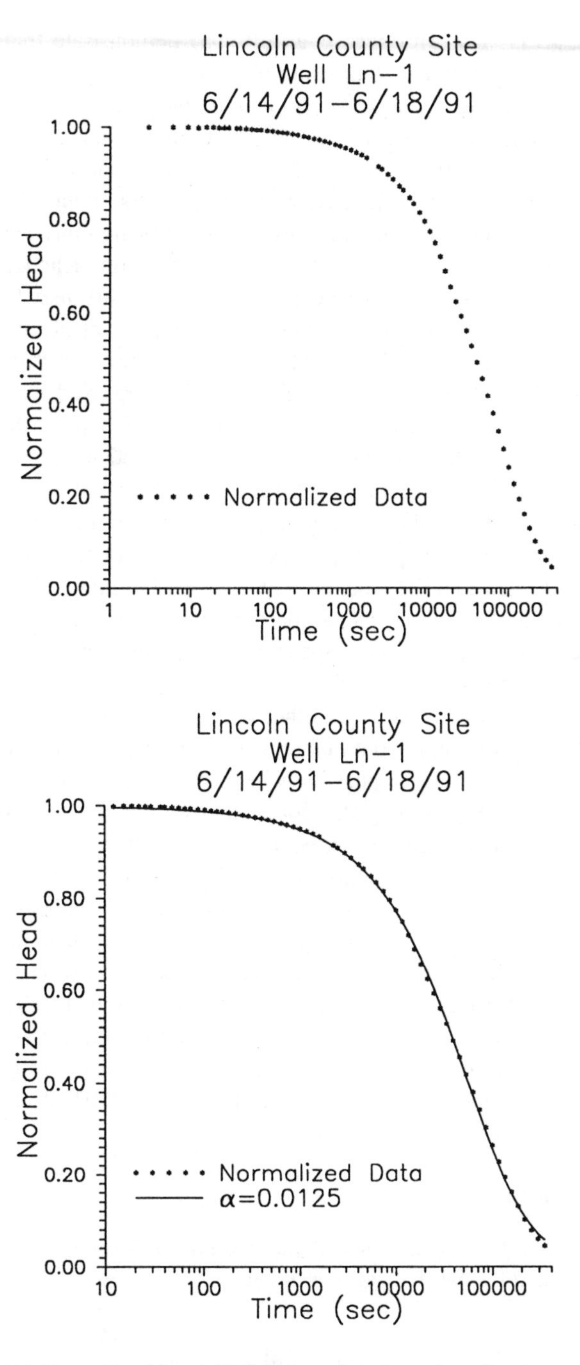

FIGURE 5.4 (A) Normalized head ($H(t)/H_0$) vs. log time plot of a slug test performed in well Ln-1 at the Lincoln County, Kansas monitoring site; (B) Normalized head vs. log time plot and the best-fit Cooper et al. type curve.

Papadopulos et al. (1973). McElwee et al. (1995a) demonstrate that reliable estimates of α will be difficult to obtain using the density and quality of data that are normally collected during a slug test. Fortunately, however, this insensitivity to α does not translate into a large error in the conductivity estimate. Papadopulos et al. (1973) report that for $\alpha < 10^{-5}$, a two order of magnitude error in the α estimate will translate into an error in the conductivity estimate of less than 30%. At larger α (10^{-5} to 10^{-1}), differences in the shapes of the type curves are more readily apparent; so, the potential for significant errors in the α estimate is diminished. However, even when a reasonable estimate for α has been obtained, there still may be considerable uncertainty regarding the specific storage estimate as a result of uncertainty about the effective screen radius. Butler et al. (1996) have pointed out that the quality of the estimates for both α and specific storage can be improved when observation wells are used, an issue that will be discussed further in Chapter 10.

The standard application of the Cooper et al. method uses a match between normalized head data and theoretical type curves to estimate the hydraulic conductivity and specific storage of the formation. As discussed in the previous paragraph, the uncertainty regarding the specific storage estimate is often quite large. Several authors have suggested that the Cooper et al. method be modified so that both the normalized head and the logarithmic temporal derivative of the normalized head be matched by theoretical relationships. Spane and Wurstner (1993) show that plots of the logarithmic temporal derivative of the normalized head can provide useful information that may not be obtainable by other means. Ostrowski and Kloska (1989) and Mishra (1997) have pointed out that reductions in the uncertainty regarding the specific storage estimate may be possible when the derivative data are also used in the matching process. Figure 5.5 is a plot of the normalized head type curves of Figure 5.2 supplemented with the corresponding derivative type curves. Note that the differences in the shape and amplitude of the derivative curves become quite small when α gets below 1.e-4. Unfortunately, in practical applications, it may be difficult to exploit these differences to significantly reduce the uncertainty regarding the specific storage estimate. This approach is most effective when H_0 is large and high-accuracy pressure transducers are used. Spane and Wurstner (1993) describe use of the DERIV program (see Appendix A for availability information) to calculate the derivative data and discuss various issues regarding those calculations. Horne (1995) presents a similar overview of the derivative calculation from a petroleum-engineering perspective.

The Hvorslev Method

The Hvorslev method for the analysis of slug tests in fully penetrating wells is based on the mathematical model defined as follows:

$$\frac{\partial^2 h}{\partial r^2} + \frac{1}{r}\frac{\partial h}{\partial r} = 0 \quad (5.5a)$$

$$H(0) = H_0 \quad (5.5b)$$

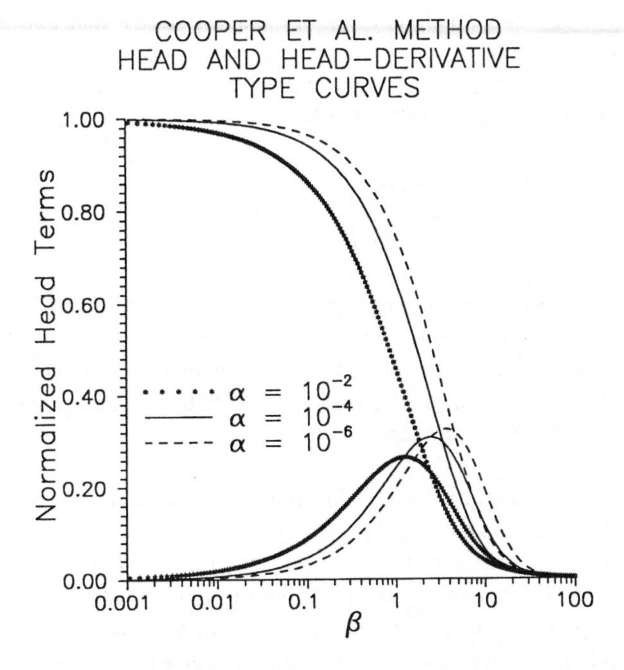

FIGURE 5.5 Normalized head ($H(t)/H_0$) (upper set) and logarithmic temporal derivative of normalized head $\left(\frac{d(H(t)/H_0)}{d(\ln t)}\right)$ (lower set) vs. logarithm of β type curves generated with the Cooper et al. model.

$$h(R_e, t) = 0, t > 0 \tag{5.5c}$$

$$h(r_w, t) = H(t), t > 0 \tag{5.5d}$$

$$2\pi r_w K_r B \frac{\partial h(r_w, t)}{\partial r} = \pi r_c^2 \frac{dH(t)}{dt}, t > 0 \tag{5.5e}$$

where: R_e = effective radius of the slug test, [L].

The mathematical model defined by Equations (5.5a) to (5.5e) differs from that on which the Cooper et al. method is based in three critical aspects. First, although the hydraulic head in the formation still varies with time in response to head changes in the well, the right-hand side of Equation (5.5a) is zero because the specific storage is assumed to be so small that its effects can be neglected. Thus, any change in head in the well is instantly propagated throughout the flow system, a situation designated as a quasi-steady-state representation of the slug-induced flow. Second, the slug does not necessarily have to be introduced in an instantaneous fashion. Third, lateral constant-head boundaries are at a finite distance (R_e) from the test well. Note that the last two differences are a direct result of the quasi-steady-state conceptualization of the slug-induced flow.

The analytical solution to the mathematical model defined by Equations (5.5a) to (5.5e) can be written as (Chirlin, 1989):

$$\ln\left(\frac{H(t)}{H_0}\right) = -\frac{2K_r Bt}{r_c^2 \ln(R_e/r_w)} \tag{5.6}$$

An important characteristic of Equation (5.6) is that a plot of this solution in the format of the logarithm of normalized head vs. time is a straight line. The Hvorslev method involves calculating the slope of that straight line and using that value to estimate the radial component of hydraulic conductivity. The method essentially consists of the following four steps:

1. The logarithm of the normalized response data is plotted vs. the time since the test began as shown in Figure 5.6A;
2. A straight line is fit to the data plot either via visual inspection or an automated regression routine as shown in Figure 5.6B;
3. The slope of the fitted line is calculated. A very common method for calculating the slope is to estimate the time at which a normalized head of 0.368 (the natural logarithm of which is –1) is obtained. This time is defined as the basic time lag according to the terminology of Hvorslev (1951) and is designated as T_0. Since the logarithmic head term and the time term are both zero at the start of the test, the slope determined in this manner is just $\log_{10}$ of 0.368 over T_0, which, when written in terms of the natural logarithm, becomes $-1/T_0$;
4. The radial component of hydraulic conductivity can be estimated by rewriting Equation (5.6) in terms of the slope calculated using a normalized head of 0.368:

$$K_r = \frac{r_c^2 \ln(R_e/r_w)}{2BT_0} \tag{5.7}$$

where: T_0 = time at which a normalized head of 0.368 (commonly rounded to 0.37) is obtained, [T].

There are three issues of practical importance with respect to the Hvorslev method. First, Equation (5.7) requires an estimate of R_e, which has been defined as the effective radius of the slug test. Although the definition of R_e may indicate to the contrary, this quantity should be viewed as an empirical parameter and not as the actual effective radius of a slug test, which, in reality, will be a function of α as discussed further in Chapter 10. Most past work on slug tests has recommended that either the length of the well screen (an extension of the approach used for partially penetrating wells) or a distance 200 times the effective radius of the well screen

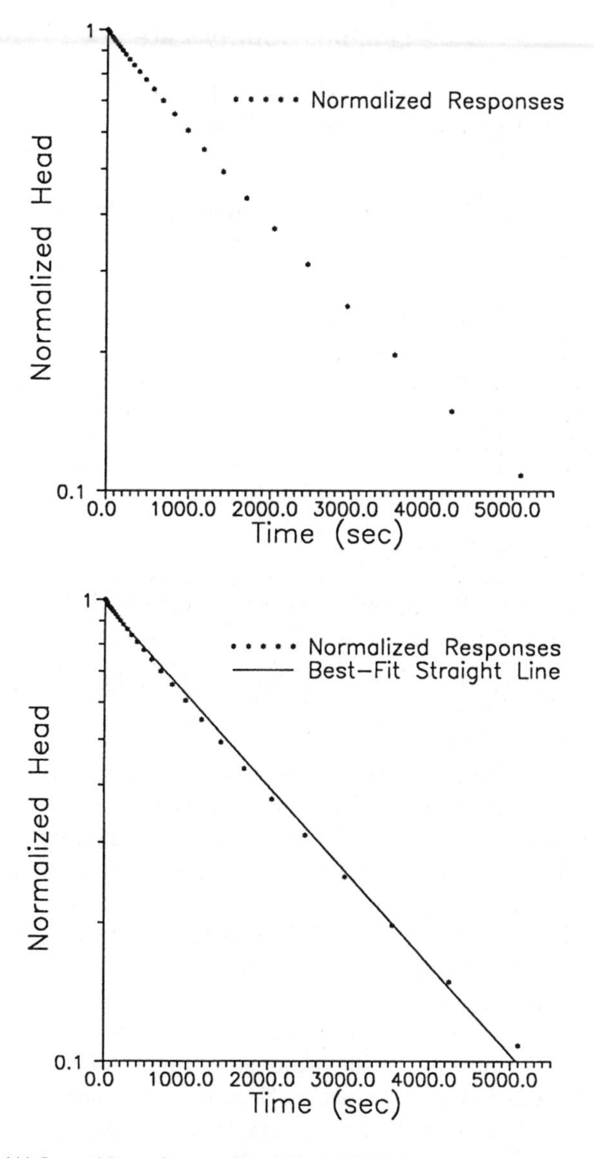

FIGURE 5.6 (A) Logarithm of normalized head ($H(t)/H_0$) vs. time plot simulated with the solution of Cooper et al.; (B) Example of the fitting procedure used in the Hvorslev method.

(U.S. Department of Navy, 1961) be used for R_e. The latter recommendation, a distance 200 times the effective radius of the well screen, is used here. Since a fully penetrating well commonly has a screen length on the order of tens to hundreds of times the screen radius and R_e only appears within a logarithmic term, the difference between these two approaches is relatively small in most cases.

A second issue of practical importance is that of fitting a straight line to the semilog plot of the test data. The Hvorslev method is based on the assumption that

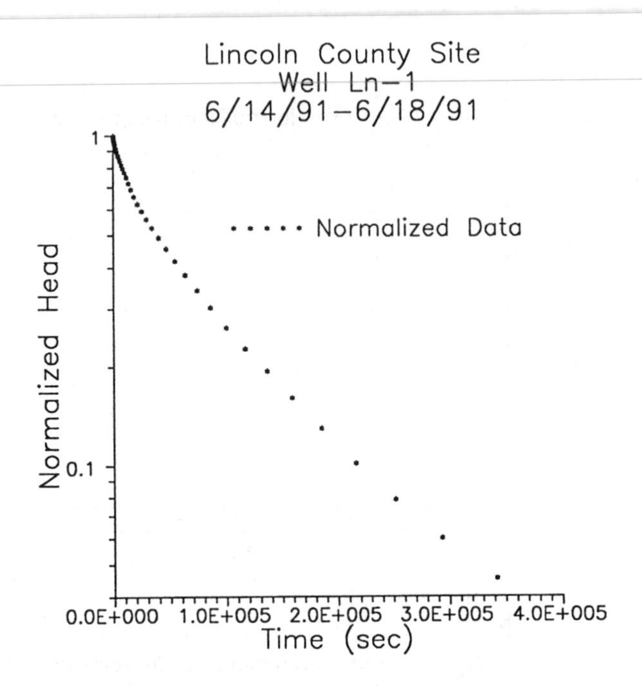

FIGURE 5.7 Logarithm of normalized head ($H(t)/H_0$) vs. time plot of a slug test performed in well Ln-1 at the Lincoln County, Kansas monitoring site.

a plot of the logarithm of the normalized response data vs. time will be linear. When the effect of the elastic storage mechanisms can be neglected, Equation (5.6) shows that this assumption is appropriate. It is not uncommon, however, for the storage mechanisms to have some effect on the response data. In such cases, test data will display a distinct concave-upward curvature when plotted in a semilog format (e.g., Figure 5.7). Chirlin (1989) provides a theoretical explanation for this behavior, demonstrating that the concave-upward curvature is primarily a function of the dimensionless storage parameter (α).

When a graph of the response data has a pronounced concave-upward curvature, there may be considerable uncertainty about how to fit a straight line to the plotted data. Figures 5.8A to 5.8C demonstrate the three most common approaches that are used to fit a straight line to data plotted in the Hvorslev format. Figure 5.8A illustrates one of the most common fitting approaches, which emphasizes the early-time portions of the data record. This is the type of fit that is usually obtained with an automated analysis procedure, because most of the available software for the analysis of slug-test data performs the straight-line fit using the actual response data and not the logarithm of those data. Figure 5.8B demonstrates a frequently used approach that has developed from similar techniques for the analysis of tests in unconfined formations (Bouwer, 1989). In this case, the concave-upward curve is represented as two straight line segments, with the fitting being performed on the second linear segment. Figure 5.8C depicts a final approach in which an attempt is made to fit a straight line to the entire data plot. This can be the result of a manual fit to the entire data set or the use of an automated analysis procedure based on the logarithm of the response data.

Although all three approaches are commonly used in practice, there has been little attempt to assess which is most appropriate. Butler (1996) reports the results of a study that found the best approach is to fit a straight line to normalized response data in the range of 0.15 to 0.25 (e.g., Figure 5.8B). This finding is in keeping with the double-straight-line theory of Bouwer (1989), which will be discussed at length in Chapter 6.

A final issue of practical importance is that of the relative speed of slug introduction. The Hvorslev method, unlike that of Cooper et al., does not require that the slug be introduced in a near-instantaneous fashion relative to the formation response. In fact, as a result of the quasi-steady-state representation of slug-induced flow, there is no assumption about the relative speed of slug introduction in the underlying mathematical model. The only assumption in this regard is that slug introduction has been completed prior to the collection of response data (Equations [5.5b] and [5.5e]). In formations of very high hydraulic conductivity, this assumption may equate to near-instantaneous introduction if much data are to be collected before heads have returned to static levels. That, however, is not true in the general case. Thus, the Hvorslev method can often be used to analyze response data that have been affected by noninstantaneous slug introduction. In these cases, the translation method of Pandit and Miner (1986) discussed in Chapter 4 should be employed prior to the performance of the actual analysis. It is important to emphasize again, however, that the Hvorslev method is dependent on the response data being unaffected by the storage properties of the well-formation configuration (i.e., the α parameter). In cases where α is influencing test responses, non-instantaneous slug introduction will introduce error into the hydraulic conductivity estimate. In those situations, every effort should be made to ensure that the slug is introduced in as rapid a manner as possible. As discussed earlier, a pronounced concave-upward curvature to a plot of the logarithm of head vs. time is often an indication that the α parameter is having a significant effect on test responses.

The same slug test used to demonstrate the Cooper et al. method can also be employed to illustrate an application of the Hvorslev procedure. Figure 5.7 is a plot of the test data in a logarithm of normalized head vs. time format. The concave-upward curvature of the plot clearly indicates that the storage parameter is affecting test responses. Figure 5.8B displays a straight line fit to the normalized head range recommended by Butler (1996). Substitution of the T_0 estimate (120,974 sec) obtained for this fit into Equation (5.7) yields a hydraulic conductivity estimate of 3.77×10^{-4} m/d, a value that is approximately 8% higher than the estimate obtained with the Cooper et al. method. The close agreement between these estimates is a further demonstration of the appropriateness of the recommendations given in Butler (1996). Note that if an approach similar to that illustrated in Figure 5.8A is used, an estimate greater than twice that of the Cooper et al. value will be obtained. Pursuing an approach similar to that shown in Figure 5.8C will yield an estimate that is 30% higher than the Cooper et al. value.

The estimation of T_0 for the fitting procedure depicted in Figure 5.8B requires some explanation. The T_0 estimate in this case is with respect to the head value at

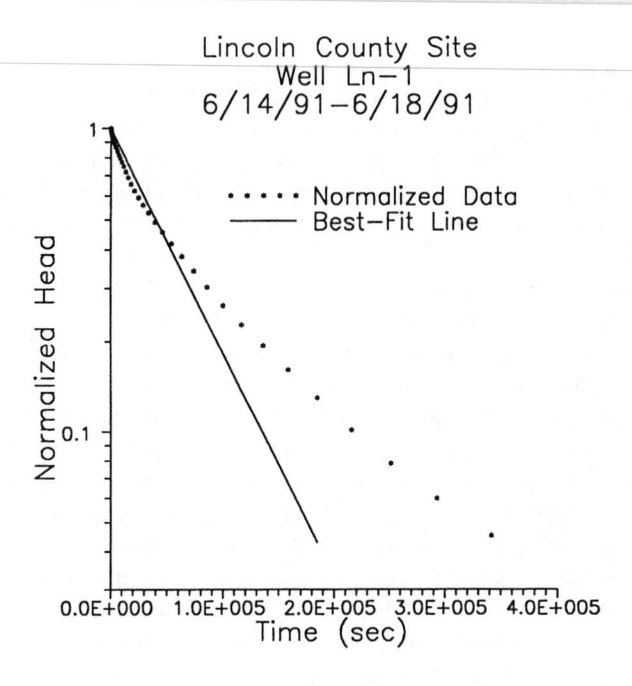

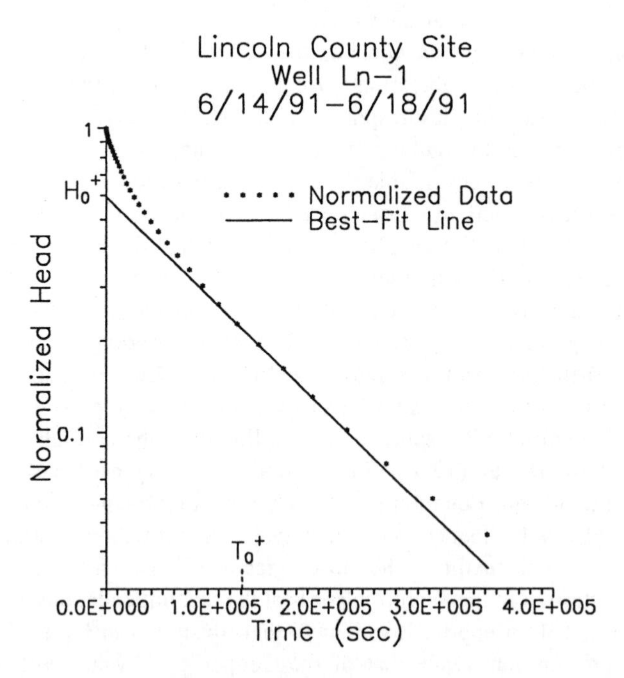

FIGURE 5.8 Logarithm of normalized head ($H(t)/H_0$) vs. time plots illustrating the three most common approaches used in the Hvorslev method for fitting a straight line to test data (details of the procedures used in A, B, and C are given in text; H_0^+ and T_0^+ defined in text).

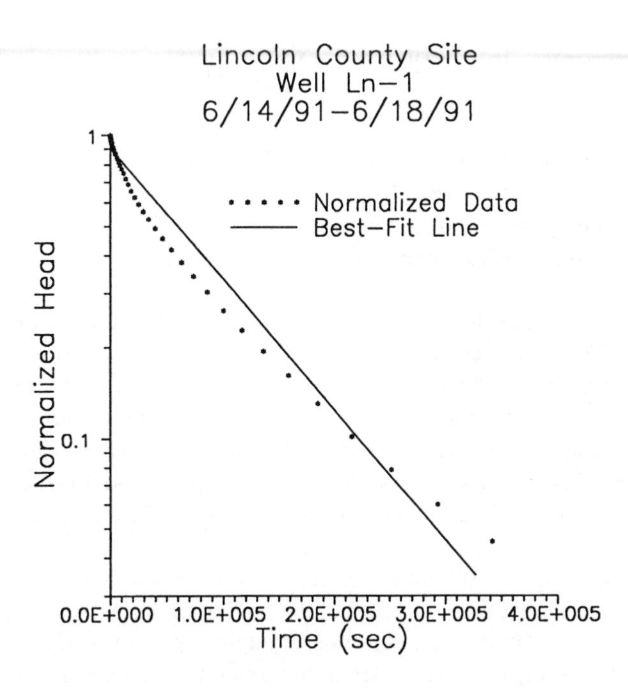

FIGURE 5.8 (continued)

which the fitted line intersects the y axis (H_0^+ in Figure 5.8B), i.e., it is the time at which a value of 0.368 is obtained for the normalized head defined as $H(t)/H_0/H_0^+$. The T_0 estimate obtained using this approach is designated as T_0^+ in Figure 5.8B.

The Peres et al. Approximate Deconvolution Method

The techniques commonly employed for the analysis of slug tests performed as part of shallow groundwater investigations estimate the hydraulic conductivity of the formation using various normalized plots of the response data. The approximate deconvolution method of Peres et al. (1989), however, is based on a completely different approach. Instead of working directly with the response data, this technique is based on transforming the response data into the equivalent drawdown that would be produced by a constant rate of pumping at the test well in the absence of wellbore storage effects. The equivalent drawdown data are then analyzed using the Cooper-Jacob semilog straight-line method for constant-rate pumping tests (Cooper and Jacob, 1946; Kruseman and de Ridder, 1990) to obtain an estimate of the hydraulic conductivity of the formation.

The approximate deconvolution method is based on a theoretical relationship between slug-test responses and pumping-induced drawdown that was originally derived for the fully penetrating well case by Ramey and Agarwal (1972). Those authors found that slug-test responses and drawdown produced by a constant rate of pumpage can be related, in the case of no well skin, by the following dimensionless equation:

$$\frac{H(\tau,\alpha)}{H_0} = \frac{1}{\alpha}\frac{ds_D}{d\tau}(\tau,\alpha) \tag{5.8}$$

where: s_D = dimensionless drawdown = $(4\pi K_r Bs/Q)$;
s = actual drawdown, [L];
Q = pumpage rate from test well, $[L^3/T]$;
τ = dimensionless time = $4\beta/\alpha$.

Peres et al. (1989) demonstrate that this relationship will hold for any well-formation configuration. Thus, slug-test type curves for any given well-formation configuration can be readily generated from the solution for drawdown produced by a constant rate of pumping in that same configuration. It is important to emphasize that the configurations must be the same. Thus, if the effect of wellbore storage is to be included in the slug-test type curves, the corresponding constant-rate pumping case must incorporate the effects of wellbore storage (e.g., Papadopulos and Cooper, 1967).

Since most analytical solutions for slug tests in well-formation configurations of practical interest have been developed, the finding of a correspondence between solutions for slug- and pumping-test responses might appear to be of limited practical significance. However, Peres et al. demonstrate that the real advantage of the relationship given by Equation (5.8) is that it can be used to actually transform response data from slug tests into the equivalent drawdown that would be produced by a constant rate of pumpage. These equivalent drawdown data can then be analyzed using standard methods for the analysis of constant-rate pumping tests.

The first step in the approximate deconvolution method is to transform the response data into the equivalent drawdown data for the same well-formation configuration. The equation for this transformation can be readily obtained by integrating both sides of Equation (5.8) from 0 to τ_D:

$$s_D(\tau_D,\alpha) = \alpha\int_0^{\tau_D}\frac{H(\tau,\alpha)}{H_0}\,d\tau \tag{5.9a}$$

An expression for the equivalent drawdown derivative data can be obtained by multiplying both sides of Equation (5.8) by $\tau\alpha$:

$$\frac{ds_D(\tau,\alpha)}{d\ln\tau} = \tau\alpha\frac{H(\tau,\alpha)}{H_0} \tag{5.9b}$$

Equations (5.9a) and (5.9b) allow the equivalent drawdown and drawdown derivative data for a particular well-formation configuration to be generated using the response data from a slug test in the same configuration. Peres et al. (1989) propose that these equivalent drawdown and drawdown derivative data be analyzed with type curve methods developed by Bourdet et al. (1983) and Onur and Reynolds (1988).

These methods use both drawdown and drawdown derivative data in an attempt to reduce the uncertainty produced by the nonuniqueness of drawdown type curves. However, these methods are not particularly effective for most slug tests performed as part of shallow groundwater investigations because wellbore storage effects dominate the equivalent drawdown and drawdown-derivative type curves. Thus, there may be considerable uncertainty about the conductivity estimate as a result of the influence of wellbore storage.

The major effect of wellbore storage in a constant-rate pumping test is to cause the rate of flow from the formation into the well (the "sand-face discharge" in the terminology of petroleum engineering) to vary with time. Although the pumping rate is constant, the rate of flow from the formation into the well increases with time as the amount of water removed from storage in the well decreases. Relatively little can be learned about the formation during the period in which removal of water from storage in the well is a dominant mechanism because wellbore storage effects are masking the impact of formation properties. In an attempt to get more information about formation properties from drawdown during this period, petroleum engineers have developed deconvolution approaches based on Duhamel's theorem (Carslaw and Jaeger, 1959). These approaches, however, are computationally intensive and are prone to numerical instabilities, thereby limiting their use in field applications (Horne, 1995). Instead, approximate deconvolution approaches are normally used. Gladfelter et al. (1955) were the first to propose an approximate deconvolution approach known as the rate normalization approach in petroleum engineering. Aron and Scott (1965) introduced an analogous approach in the groundwater literature. Ramey (1976) demonstrates that the rate-normalization approach produces a very reasonable approximation of system behavior.

The rate normalization approach is based on normalizing the drawdown in the test well by the flow rate from the formation at that same time. This normalized drawdown is approximately equal to the drawdown that would be found in the case of negligible wellbore storage, i.e., the conditions during which the finite-radius well representation of Hantush (1964) is appropriate (Ramey, 1976). This approach is particularly useful for times at which the Cooper-Jacob approximation of the Theis equation is applicable, i.e., at dimensionless times (τ) greater than 100. Under these conditions, the rate normalization approach can be written as:

$$\frac{s_D(\tau,\alpha)}{q_D(\tau)} \approx \tilde{s}_D \tag{5.10}$$

where: q_D = dimensionless flow rate from the formation;
$\tilde{s}_D$ = dimensionless drawdown for case of negligible wellbore storage.

For the case of a slug test, the dimensionless flow rate can be written as:

$$q_D = 1 - \left(ds_D/d\tau\right)/\alpha \tag{5.11a}$$

Substituting this expression into Equation (5.10) produces the form of the rate normalization approximation used for slug tests (Peres et al. [1989]):

$$\frac{s_D(\tau,\alpha)}{1-\left(ds_D/d\tau\right)/\alpha} \approx \tilde{s}_D \tag{5.11b}$$

The big advantage of working with $\tilde{s}_D$ is that the Cooper-Jacob semilog approach can be employed at a much smaller time than in the case when wellbore storage effects are influencing drawdown. Thus, by removing the effect of wellbore storage, the Cooper-Jacob method can be used to analyze equivalent drawdown data over the range of normalized heads that are usually obtained during a slug test performed for environmental applications.

The Peres et al. approximate deconvolution method essentially consists of the following five steps:

1. The equivalent drawdown and drawdown derivative data are calculated from the response data using Equations (5.9a) and (5.9b). These equations can be written in terms of dimensional quantities as:

$$s = \int_0^t \frac{H(\xi)}{H_0} d\xi \tag{5.12a}$$

and

$$\frac{ds}{d\ln t} = t\frac{H(t)}{H_0} \tag{5.12b}$$

respectively,

where: s = equivalent drawdown for case of a constant pumping rate in a finite radius well with nonnegligible wellbore storage, [T].

2. The dimensionless equivalent drawdown for the case of negligible wellbore storage is calculated using Equation (5.11b), which can be written in terms of the equivalent drawdown of Equations (5.12a) and (5.12b) as:

$$\tilde{s}_D \approx \left(\frac{4K_rB}{r_c^2}\right)\frac{s}{1-\frac{1}{t}\left(\frac{ds}{d\ln t}\right)} \tag{5.13a}$$

For plotting purposes, Equation (5.13a) can be rewritten in terms of the dimensional equivalent drawdown as:

$$\tilde{s} = \left(\frac{r_c^2}{4K_rB}\right)\tilde{s}_D \approx \frac{s}{1-\frac{1}{t}\left(\frac{ds}{d\ln t}\right)} \tag{5.13b}$$

where: $\tilde{s}$ = equivalent drawdown for the case of negligible wellbore storage, [T];

3. The equivalent drawdown for the case of negligible wellbore storage ($\tilde{s}$) is plotted vs. the logarithm of time since the test began;
4. A straight line is fit to the data plot either via visual inspection or an automated regression routine, emphasizing data at moderate to large times;
5. The slope of the straight line is calculated and the radial component of hydraulic conductivity is estimated by substituting the Cooper-Jacob approximation (Kruseman and de Ridder, 1990) for $\tilde{s}_D$ into the middle expression of Equation (5.13b):

$$\tilde{s} \approx \frac{2.30r_c^2}{4K_rB}\log\left(\frac{4K_rt}{S_sr_w^2}\right) \quad (5.14a)$$

If the slope is calculated as the change in $\tilde{s}$ over a log cycle in time ($\Delta\tilde{s}$), then the equation for estimation of the radial component of hydraulic conductivity can be written as:

$$K_r = \frac{2.30r_c^2}{4B\Delta\tilde{s}} \quad (5.14b)$$

Note that an estimate for S_s can also be obtained from the intercept of the straight line. This estimate, however, will be extremely sensitive to nonideal conditions in the immediate vicinity of the well; so, relatively little significance can be attached to its magnitude (e.g., Butler, 1988).

Since the approximate deconvolution method is relatively unknown among the hydrologic community, both a hypothetical example and a field demonstration of the approach will be given to supplement this presentation. Figure 5.9A is a plot of simulated responses vs. the logarithm of time since test initiation for the case of a well fully screened across a confined aquifer. Figure 5.9B is a log-log plot of the equivalent drawdown and drawdown derivative terms calculated from the simulated responses using Equations (5.12a) and (5.12b). Note that these equivalent drawdown terms were calculated assuming that noise-free responses were available to a normalized head of 0.0001. Although noise-free data at such low normalized heads are virtually impossible to obtain in shallow groundwater applications, curve matching techniques based on the derivative plots are extremely useful if such data are available. The reason for this is that the approximate position of the horizontal line towards which the derivative curve converges can be identified, considerably reducing the uncertainty in the type-curve match (e.g., Horne, 1995). Figure 5.9C is a semilog plot of the equivalent drawdown calculated from the simulated responses using Equation (5.13b). If the slope of the straight line labelled Best-Fit Line A is substituted into Equation (5.14b), a hydraulic conductivity estimate 0.4% lower than the value used to generate the simulated responses is obtained. Note that if relatively noise-free data are only available until a normalized head of approximately 0.01

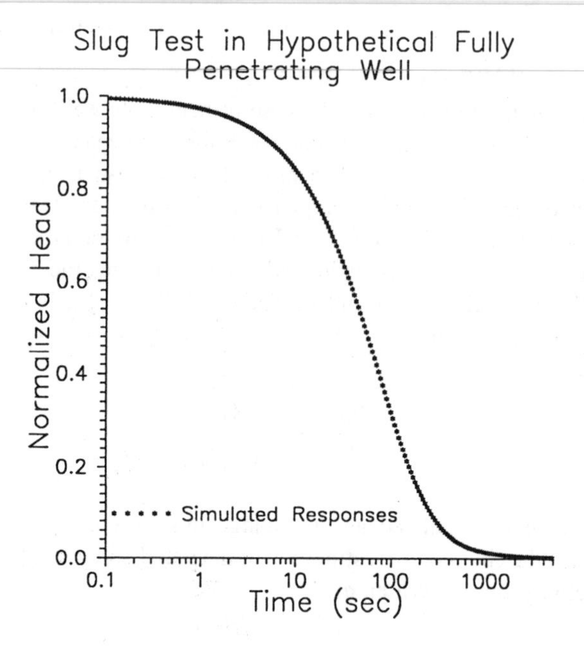

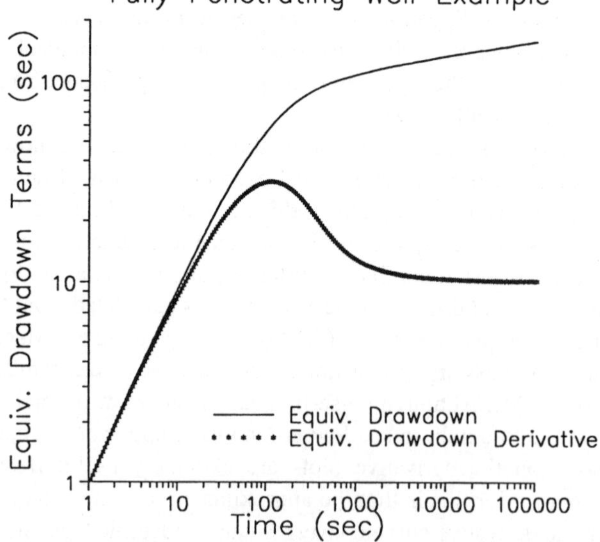

FIGURE 5.9 (A) Normalized head ($H(t)/H_0$) vs. log time plot simulated with the solution of Cooper et al.; (B) Equivalent drawdown and drawdown derivative (prior to removal of wellbore storage effects) plots for simulated responses of Figure 5.9A; (C) Equivalent drawdown (after removal of wellbore storage effects) plot for simulated responses of Figure 5.9A (Definitions of equivalent drawdown and drawdown derivative terms are given in text.)

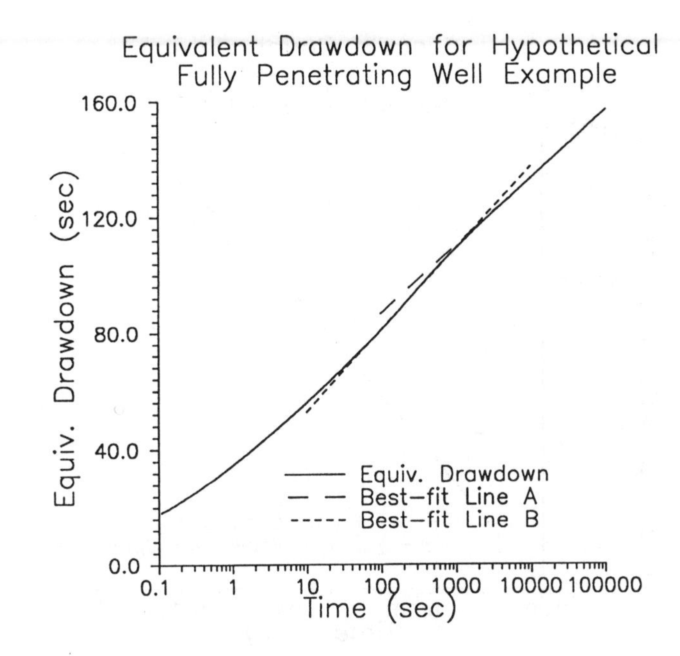

FIGURE 5.9 (continued)

(Best-Fit Line B), a level possible in many groundwater applications, the hydraulic conductivity estimate obtained from Equation (5.14b) will underpredict the actual value by 18%.

The same slug test used to demonstrate the Cooper et al. and Hvorslev methods can be employed to illustrate a field application of the approximate deconvolution method. Figure 5.4A is a plot of the response data in a normalized head vs. log time format. Figure 5.10A is a log-log plot of the equivalent drawdown and drawdown derivative terms calculated from the response data using Equations (5.12a) and (5.12b). A comparison of Figures 5.10A and 5.9B shows that the plot of the derivatives calculated from the field data does not extend for a sufficient period of time to prevent ambiguity from being introduced into estimates obtained from curve matching techniques. In this test, data collection was terminated at a normalized head of 0.045. For effective application of derivative-based curve matching approaches, normalized heads considerably below 0.01 are needed. Figure 5.10B is a semilog plot of the equivalent drawdown calculated from the response data using Equation (5.13b). If the slope of the best-fit straight line is substituted into Equation (5.14b) along with information from Table 5.1, a hydraulic conductivity estimate of 3.58×10^{-4} m/d is obtained, a value within 3% of the value calculated using the Cooper et al. method. Note that only equivalent drawdowns from the latter portion of the plot (corresponding to normalized responses of less than 0.30) were used for this fit in keeping with the results obtained from the hypothetical example.

There are a number of issues of practical importance with respect to the approximate deconvolution method. First, the approach requires that the response data be transformed into an equivalent drawdown (Equation [5.13b]). Spane and Wurstner

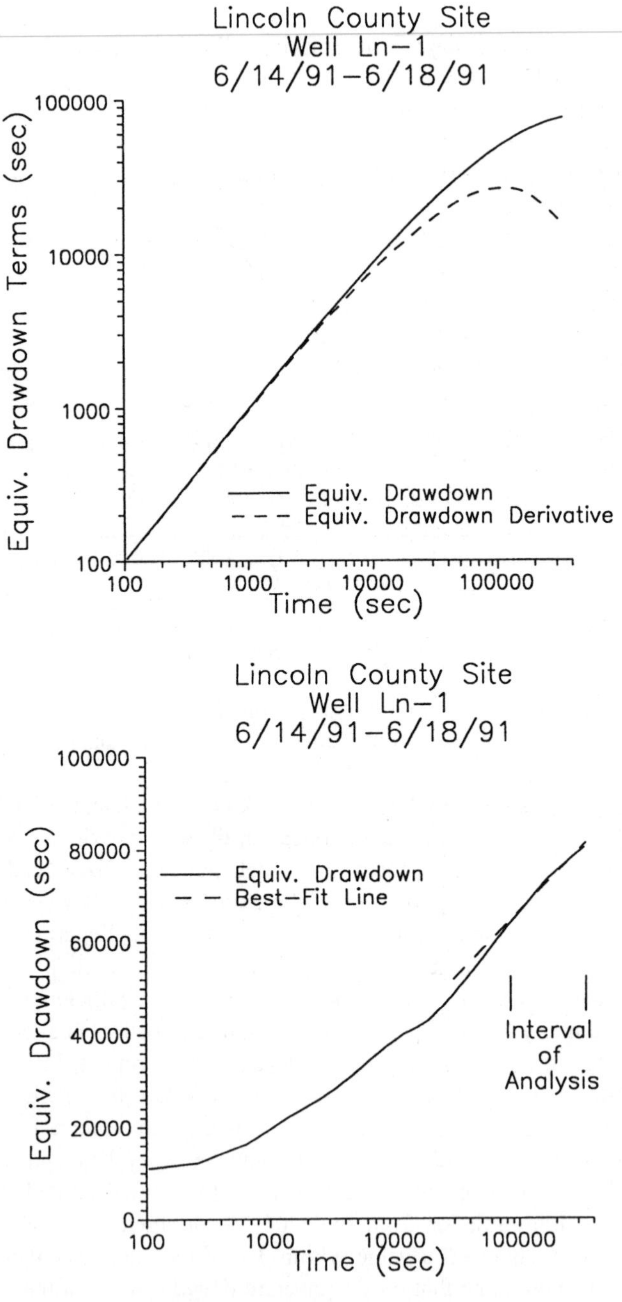

FIGURE 5.10 (A) Logarithm of equivalent drawdown and drawdown derivative (prior to removal of wellbore storage effects) vs. logarithm of time plot of a slug test performed in well Ln-1 at the Lincoln County, Kansas monitoring site; (B) Equivalent drawdown (after removal of wellbore storage effects) vs. log time plot of test at well Ln-1 (definitions of equivalent drawdown and drawdown derivative terms are given in text; interval denoted as "Interval of Analysis" indicates the portion of the response data used in the analysis).

(1993) present a program, DERIV, that can be used to calculate the equivalent drawdown and drawdown derivative data for the case of nonnegligible wellbore storage (Equations [5.12a] and [5.12b]). This program can be readily modified or combined with a spreadsheet program to perform the calculations required in Equation (5.13b). All examples of the approximate deconvolution method given in this book were performed with the DERIV program (see Appendix A for information on program availability).

Second, the method is most effective when the recovery data are closely spaced in time and relatively noise-free data are available at small normalized heads. Data sets in which the normalized head changes by more than a few percent between adjacent measurements can be more difficult to analyze, especially when noise introduced by measurement and acquisition equipment is relatively large. The technique works best when H_0 is large, high accuracy pressure transducers are available, background noise is quite small, and the test is run to complete recovery.

Even under the most ideal conditions, the approximation of the deconvolution operation incorporated in the rate normalization approach can introduce error into the conductivity estimate. Ramey (1976) assesses the error that is introduced into the calculated slope, and thus the conductivity estimate, as a result of this approximation. He found that this error is only of significance over a relatively narrow interval of time where the slope of the equivalent drawdown plot is greater than expected. In the worst case (i.e., when only data from that interval are used in the analysis), the hydraulic conductivity of the formation will be underpredicted by about 26%. This result is in agreement with the 18% underprediction obtained from best-fit line B in Figure 5.9C, which extends over a time range that includes this interval. In field applications, however, this interval will be difficult to identify and is of very limited practical significance.

Finally, the major advantage of the approximate deconvolution method is that it allows the Cooper-Jacob method to be applied to the analysis of response data from slug tests. As Butler (1988,1990) has emphasized, a conductivity estimate obtained using the Cooper-Jacob semilog method is independent of nonideal conditions in the immediate vicinity of the pumping well. As will be discussed further in Chapter 9, this characteristic of the Cooper-Jacob method makes the approximate deconvolution approach potentially of great use for tests performed in the presence of a low-permeability well skin. Given the importance of diminishing the impact of nonideal conditions in the immediate vicinity of the well, the straight line should always be fit to the latter portion of the equivalent drawdown plot.

Additional Methods

The vast majority of slug tests in fully penetrating wells in confined formations are analyzed using one of the previously described methods. However, there are several other approaches that have been periodically employed in the literature for this same purpose. The two most common are the method of Ferris and Knowles (1963) and that of Dax (1987), both of which are briefly described in the following paragraphs.

The method of Ferris and Knowles was developed in the early 1950s, although it was not described in a widely available publication until much later (Ferris and

Knowles, 1963). References to the method have appeared periodically in the literature over the last three decades (e.g., Leap, 1984; Campbell et al., 1990). The mathematical model underlying the Ferris and Knowles technique is similar to that of the Cooper et al. method with the exception that the effect of well-bore storage in the test well is assumed to be negligible. The practical ramification of this assumption is that the technique is only applicable to normalized heads less than about 0.0025 (Cooper et al., 1967), i.e., the very small deviations from static that are measured at the tail end of a slug test. The analytical solution to the mathematical model of Ferris and Knowles is equivalent to the Theis (1935) solution for an infinitely small duration of pumpage and can be manipulated to obtain the following expression for the calculation of the radial component of hydraulic conductivity:

$$K_r = \left(\frac{1/t}{H(t)/H_0} \right) \frac{r_c^2}{4B} \tag{5.15}$$

Note that the term in the parenthesis is the inverse of the slope of a straight line fit to a plot of normalized head vs. the inverse of test duration.

The Ferris and Knowles method consists of calculating the slope of a straight line fit to a plot of the response data and using that slope to determine the radial component of hydraulic conductivity. The approach can be illustrated using the same slug test as employed in the previous sections. Figure 5.11 is a plot of the Lincoln County test data in the format required for the Ferris and Knowles method (normalized head vs. the inverse of test duration). The dashed line is the response predicted by the Ferris and Knowles method using the hydraulic conductivity value previously estimated from the response data with the Cooper et al. method. Note that only the last head measurement taken during the Lincoln County test (normalized head = 0.045, time = 341,640 s [≈5700 min]) hints that the slope of the data plot may be converging on the slope predicted by the Ferris and Knowles method. Unfortunately, a duration of over 80,000 min would be required in this case before the normalized heads would near the range (<0.0025) at which the Ferris and Knowles method is strictly applicable. The initial displacement (H_0) for this test was 10.35 m; so, the head value corresponding to a normalized head of 0.0025 is 0.026 m, i.e., only the last 3 cm of the recovery data would be usable. Note that a considerable underprediction of K_r will result if normalized head data in the interval A to A′ are employed to estimate a slope for use in Equation (5.15).

This example application reveals the two major limitations of the Ferris and Knowles method. First, a test must be extended for a much longer duration (well over an order of magnitude longer in this example) than that required for the Cooper et al. and Hvorslev techniques. Second, the normalized heads for which the technique is applicable are extremely small. Although in the example of Figure 5.11 an H_0 of over 10 m was employed, most slug tests for environmental applications are performed using an H_0 of less than 2 m. For these cases, the normalized head constraint translates into a deviation from static of a few millimeters, a magnitude that is on the order of/less than the background noise arising from sensors, data-acquisition

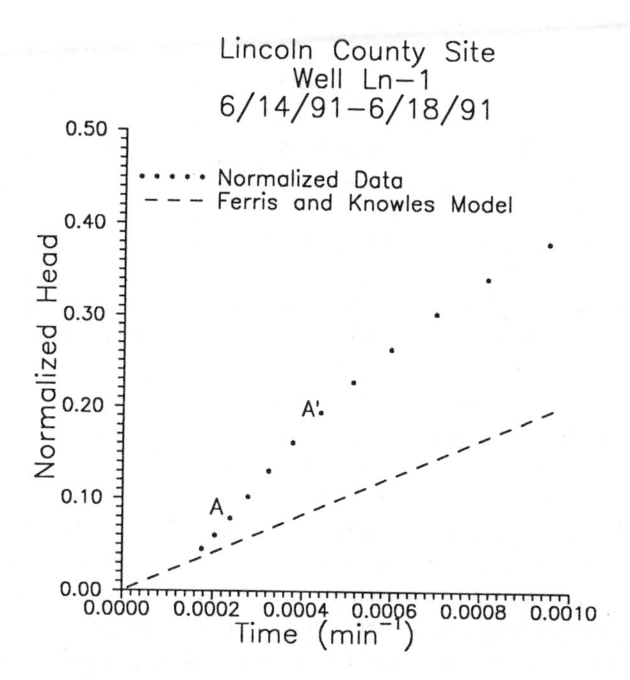

FIGURE 5.11 Normalized head ($H(t)/H_0$) vs. inverse of time plot of a slug test performed in well Ln-1 at the Lincoln County, Kansas monitoring site (A and A′ are defined in text; dashed line calculated with the K_r estimate obtained from Figure 5.4B).

devices, barometric pressure changes, pumping of nearby wells, passage of vehicles, etc. Given the severity of these two limitations, the Ferris and Knowles method should be considered of little use for the vast majority of field applications.

A second technique that has periodically been applied in the literature is the method of Dax (1987). This approach is actually quite similar to the Hvorslev method. The only real difference between the two techniques is that the Dax method employs an empirical factor (D), equivalent to the $\ln(R_e/r_w)$ term in the Hvorslev method, that is a function of the dimensionless storage parameter (α). A table of D values and an accompanying figure are provided in Dax (1987) for use in estimation of this factor. As with the Hvorslev approach, the Dax method is based on the assumption that the effect of the slug-induced disturbance is limited to a finite region in the vicinity of the test well. In the derivation of the approach, that assumption is invoked to justify the statement that at least a portion of the response data will plot as a straight line in a logarithm of normalized head vs. time format. Although the normalized head range over which the straight line fit is to be attempted is left unspecified, examination of a figure accompanying the article indicates that this interval extends from a value slightly less than 1.0 to about 0.3, and appears to be a function of α. The Dax method involves preparing a plot of the response data in a log normalized head vs. time format, and then identifying a straight line segment on that plot. The slope of that line is calculated and the radial component of hydraulic conductivity is estimated using the following expression:

$$K_r = \frac{r_c^2}{BD\lambda} \tag{5.16}$$

where: D = empirical parameter defined in Dax (1987), [dimensionless];
λ = slope of the log normalized head vs. time plot, $[T^{-1}]$.

Note that an estimate of α is required to obtain a value of D from the table/figure provided in Dax (1987).

There are two primary limitations to the Dax method. First, as with the Hvorslev method, a linear segment must be identified from the plot of the response data. When the plot has a pronounced concave-upward curvature (e.g., Figure 5.7), there may be considerable uncertainty about which portion of the response data should be represented as a linear segment. Even if a linear segment can be identified, there is no guarantee that the same segment was employed in the original calculation of the empirical factor (D). Second, as stated above, an estimate of α is required to obtain a value for D. From the tables presented in Dax (1987), it is evident that an error in K_r on the order of a factor of two to three can readily occur as a result of the use of an inappropriate α estimate. Given the magnitude of the errors that can arise as a result of use of an inappropriate data interval or D value, the Dax method should not be considered a preferred method for analysis of slug-test data. Both the Cooper et al. and Hvorslev (using the normalized head range recommended by Butler (1996)) methods should provide better estimates in virtually all situations. Note that Hinsby et al. (1992) employ both the Dax and Cooper et al. methods, among other techniques, for the analysis of a suite of slug-test data. Unfortunately, however, the Dax and Cooper et al. methods are applied to data from tests in partially penetrating wells, a situation for which neither technique strictly applies, so little insight can be gained into the relative merits of the various techniques from the results of that investigation.

Partially Penetrating Wells

Slug tests in partially penetrating wells in confined formations are primarily analyzed by one of four methods: the method of Cooper et al. (1967), that of Hvorslev (1951), the confined extensions of the method of Dagan (1978), and the KGS model (Hyder et al., 1994) and related approaches. Although originally proposed for tests in fully penetrating wells, the approximate deconvolution method of Peres et al. (1989) also has considerable potential for the analysis of tests in partially penetrating wells. These five techniques are described in the following sections, after which three less commonly utilized approaches are briefly summarized.

The Cooper et al. Method

The Cooper et al. method is based on the mathematical model defined by Equations (5.1a) to (5.1f). When applied to data from a partially penetrating well, the quantity B, formation thickness, is replaced by b, the effective screen length. The steps employed for the analysis of response data are exactly the same as those used for a fully penetrating well.

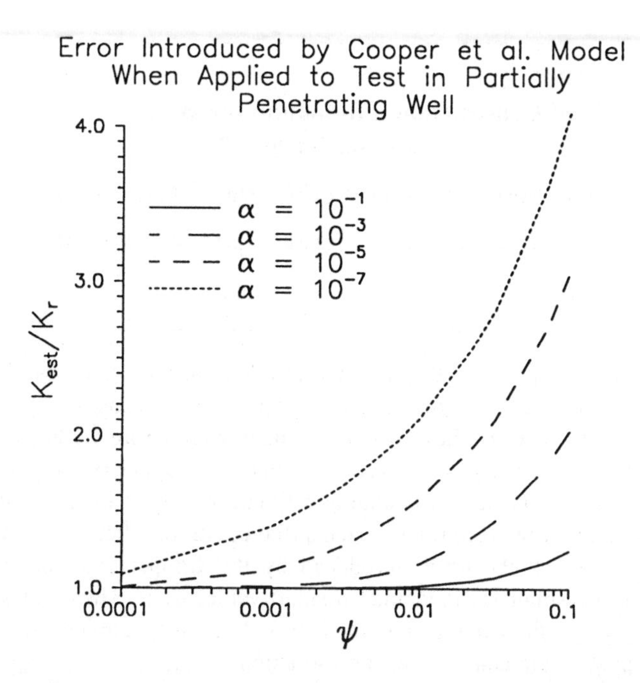

FIGURE 5.12 Plot of hydraulic conductivity ratio (Cooper et al. estimate (K_{est}) over actual conductivity (K_r)) vs. logarithm of ψ as a function of α for the case of a partially penetrating well screened at the center of a very thick formation. ψ and α are defined in text. (After Hyder et al., 1994.)

It is important to emphasize that when the Cooper et al. method is applied to test data from a partially penetrating well, the analyst has adopted a simplified representation of the flow system, i.e., there is no vertical flow in response to the slug-induced disturbance. Vertical anisotropy (vertical component of hydraulic conductivity considerably less than the horizontal component) is normally invoked to justify this simplified representation. Hyder et al. (1994) examine the ramifications of this assumption and attempt to quantify the error introduced into hydraulic conductivity estimates when the Cooper et al. model is used to analyze data from partially penetrating wells. Figure 5.12 displays the results of their investigation for a hypothetical formation in which the upper and lower boundaries are at a large distance from the screened interval. The ψ quantity plotted on the x axis is the square root of the anisotropy ratio $(K_z/K_r)^{1/2}$ divided by the aspect ratio (b/r_w) of the well. The quantity plotted on the y axis is the hydraulic conductivity estimate obtained with the Cooper et al. method (K_{est}) divided by the actual hydraulic conductivity of the hypothetical formation (K_r). A series of curves are shown for different values of the dimensionless storage parameter (α). This figure demonstrates that the error arising from the assumption of purely radial flow diminishes with decreases in ψ, as would be expected since ψ reflects the proportion of vertical to radial flow induced by the test. Decreases in ψ correspond to decreases in the anisotropy ratio or increases in the aspect ratio, the effect of either of which is to constrain the slug-induced flow to the interval bounded by the top and bottom of the well screen (i.e., the proportion

TABLE 5.3
Well Construction Information for Well 3 at Pratt County Monitoring Site 36

Well Designation	r_w(m)	r_c(m)	b(m)	B(m)	d(m)
Pratt County Site 36 Well 3	0.125	0.064	1.52	47.87	35.37

of vertical flow in response to the slug-induced disturbance decreases). Figure 5.12 also indicates that the error in the conductivity estimate decreases with increases in α. Regardless of the α value, however, it is evident from Figure 5.12 that the Cooper et al. method will always yield an estimate that is an upper bound on the actual value of K_r. The quality of that bounding estimate will be a function of ψ and α. For isotropic to slightly anisotropic systems (the condition often faced in the field), the ψ value at which the error introduced by the Cooper et al. method can be neglected is quite small for moderate to small values of α. Thus, unless α or the aspect ratio is large, the Cooper et al. method will provide estimates that are significantly greater than the actual K_r of the formation.

An example can demonstrate the use of the Cooper et al. method for the analysis of response data from a partially penetrating well. In October of 1993, a series of slug tests were performed at a monitoring well in Pratt County, Kansas (Butler et al., 1993). The well was screened in an unconsolidated alluvial sequence consisting primarily of sands and gravels with interbedded clays. Table 5.3 summarizes the well-construction information. The small aspect ratio ($b/r_w = 12$) of this well indicates that the anisotropy ratio will have to be much smaller than one for the Cooper et al. method to result in a reasonable estimate for K_r. Table 5.4 lists the test data employed in the analysis, while Figure 5.13 is a plot of that data in the normalized head vs. log time format used in the Cooper et al. method. This figure shows that the response data are characterized by a good deal of fluctuations at very early times. These

TABLE 5.4
Response Data from 10/15/93 Test #1 in Well 3 at Pratt County Monitoring Site 36

Time (s)	Head (m)	Normalized Head
0.1	0.369	0.972
0.2	0.388	1.020
0.3	0.377	0.991
0.4	0.388	1.020

TABLE 5.4 (continued)
Response Data from 10/15/93 Test #1 in Well 3 at Pratt County Monitoring Site 36

Time (s)	Head (m)	Normalized Head
0.5	0.365	0.960
0.6	0.377	0.991
0.7	0.369	0.972
0.8	0.365	0.960
0.9	0.362	0.952
1.0	0.372	0.980
1.2	0.362	0.952
1.3	0.354	0.932
1.5	0.351	0.923
1.6	0.354	0.932
1.8	0.358	0.943
2.0	0.343	0.903
2.3	0.336	0.883
2.6	0.329	0.866
2.9	0.322	0.846
3.2	0.314	0.826
3.6	0.311	0.818
4.0	0.300	0.789
4.5	0.292	0.769
5.1	0.277	0.729
5.7	0.271	0.712
6.4	0.255	0.672
7.1	0.248	0.652
8.0	0.234	0.615
9.0	0.219	0.575
10.1	0.205	0.538
11.3	0.189	0.499
12.6	0.175	0.462
14.2	0.160	0.422
15.9	0.142	0.373
17.8	0.125	0.328
20.0	0.109	0.288
22.4	0.094	0.248
25.2	0.080	0.211
28.2	0.065	0.171
31.7	0.058	0.154
35.5	0.043	0.114
39.9	0.036	0.094
44.7	0.029	0.077
50.2	0.022	0.057

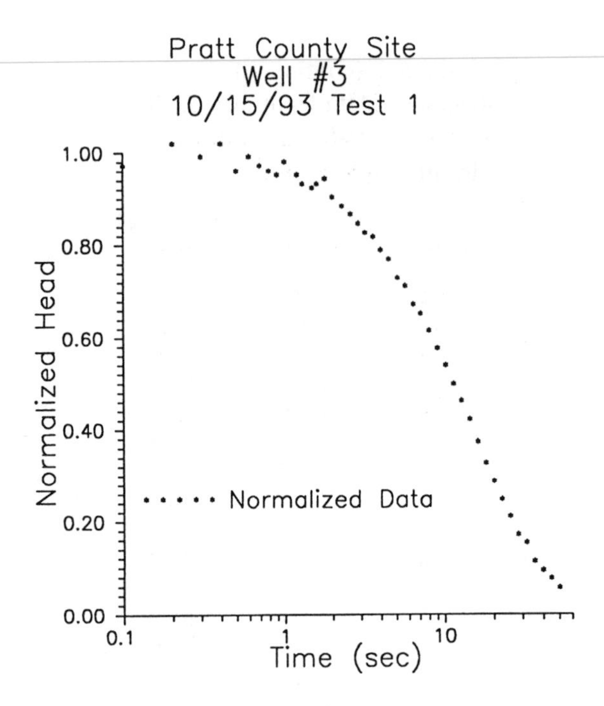

FIGURE 5.13 Normalized head ($H(t)/H_0$) vs. log time plot of a slug test performed in well 3 at a Pratt County, Kansas monitoring site.

fluctuations are related to test initiation and should be ignored when considering the quality of the match between a particular theoretical model and the test data.

Figure 5.14 displays the results of the curve-matching process used in the Cooper et al. method for α fixed at 1.90×10^{-5}, a value that was considered a reasonable lower bound on the actual α of the well-formation configuration. The systematic deviation between the normalized data and the type curve is commonly seen when applying the Cooper et al. method to analyze response data from a test in a partially penetrating well if the α parameter is constrained to the range of physical plausibility. The deviation appears to be primarily the result of a significant component of vertical flow, although it could also be the product of a low-permeability well skin as discussed in Chapter 9. The estimate of the radial component of hydraulic conductivity obtained from the analysis displayed in Figure 5.14 is 42.0 m/d, a value that must be considered a rather conservative upper bound on the K_r of the tested interval given the relationships depicted in Figure 5.12. Note that a very good match can be obtained between the normalized data and the best-fit Cooper et al. type curve using a physically implausible α ($\ll 10^{-30}$). However, little physical significance can be attached to the resulting K_r estimate; so, this approach should be avoided.

Although the Cooper et al. method will usually result in an overly conservative upper bound estimate of the K_r of the formation, an attempt should always be made to analyze data from a partially penetrating well with this approach. The absence of a systematic deviation between the test data and a type curve for a physically

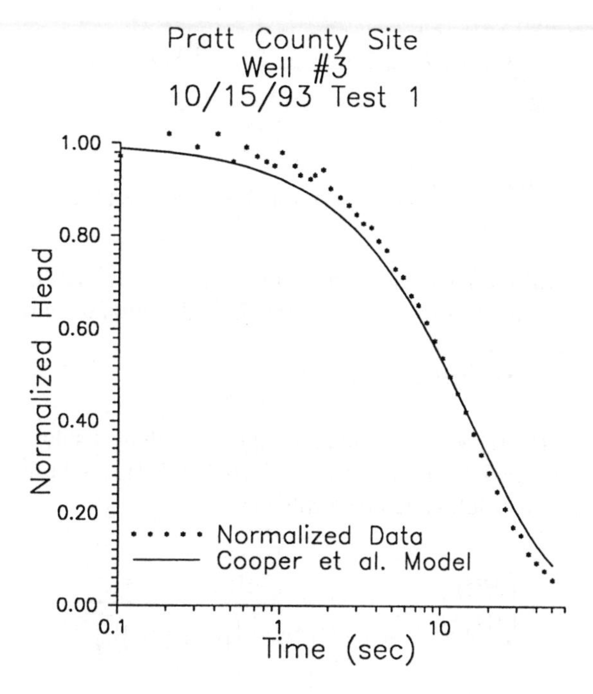

FIGURE 5.14 Normalized head vs. log time plot for the Pratt County test and a Cooper et al. type curve ($\alpha = 1.9 \times 10^{-5}$).

plausible value of α is an important indication that the vertical component of slug-induced flow has been suppressed. This suppression may be a result of a pronounced vertical anisotropy (small ψ value) or a large dimensionless storage parameter (large α value). The absence of this deviation may be one of the few clues to the significance of vertical anisotropy that can be obtained from single-well slug tests.

The Hvorslev Method

The Hvorslev method for the analysis of slug tests in partially penetrating wells is based on the mathematical model defined as follows:

$$\frac{\partial^2 h}{\partial r^2} + \frac{1}{r}\frac{\partial h}{\partial r} + \frac{K_z}{K_r}\frac{\partial^2 h}{\partial z^2} = 0 \tag{5.17a}$$

$$H(0) = H_0 \tag{5.17b}$$

$$h(\infty, z, t) = h(r, \pm\infty, t) = 0, t > 0 \tag{5.17c}$$

$$h(r_w, z, t) = H(t), d \leq z \leq (d + b), t > 0 \tag{5.17d}$$

$$2\pi r_w K_r \int_d^{d+b} \frac{\partial h(r_w, z, t)}{\partial r} dz = \pi r_c^2 \frac{dH(t)}{dt}, t > 0 \tag{5.17e}$$

$$\frac{\partial h(r_w, z, t)}{\partial r} = 0, -\infty < z < d, d + b < z < \infty, t > 0 \tag{5.17f}$$

where: K_z = vertical component of hydraulic conductivity, [L/T];
d = z position of top of screen (positive direction downward) [L];
z = vertical direction, [L];
b = effective screen length, [L].

According to Hvorslev (1951), an approximate analytical solution to the mathematical model defined by Equations (5.17a) to (5.17f) has been developed by Dachler (1936). This solution can be written as:

$$\ln\left(\frac{H(t)}{H_0}\right) = -\frac{2K_r bt}{r_c^2 \ln\left[1/(2\psi) + \left(1 + (1/(2\psi))^2\right)^{1/2}\right]} \tag{5.18}$$

As shown by Equation (5.18), a plot of the solution in the format of the logarithm of normalized head vs. time is a straight line. Thus, as in the fully penetrating case, the Hvorslev method involves calculating the slope of a straight line fit to the response data and using that value to estimate the radial component of hydraulic conductivity. The method essentially consists of the following five steps:

1. The logarithm of the normalized response data is plotted vs. the time since the test began;
2. A straight line is fit to the data plot either via visual inspection or an automated regression routine;
3. The slope of the fitted line is calculated. If the time lag (T_0) is used in this calculation as described earlier, the slope, when written in terms of the natural logarithm, becomes $-1/T_0$;
4. An estimate of the ψ parameter for the particular well-formation configuration is obtained. In most cases, a value of one is used for the anisotropy ratio;
5. The radial component of hydraulic conductivity is estimated by rewriting Equation (5.18) in terms of the slope calculated using T_0:

$$K_r = \frac{r_c^2 \ln\left[1/(2\psi) + \left(1 + (1/(2\psi))^2\right)^{1/2}\right]}{2bT_0} \tag{5.19}$$

There are several issues of practical importance with respect to the partially penetrating variant of the Hvorslev method. First, Equation (5.19) requires an estimate of the anisotropy ratio, which is incorporated in the ψ parameter. In most cases, very limited information will exist concerning the anisotropy of the material being tested. The limited data that have been reported on the vertical anisotropy of hydraulic conductivity indicate that this ratio should usually be in the range of 0.3 to 1.0, except in the case of media that consists of interbedded high-and low-conductivity materials (Freeze and Cherry, 1979). Hyder et al. (1994) examine the error introduced into K_r estimates obtained with the Hvorslev method as a result of uncertainty in the estimate of the anisotropy ratio. Their results show that this error does not exceed 10 to 20% for an anisotropy estimate within a factor of three of the actual value. Even for cases where the anisotropy ratio has been misestimated by two to three orders of magnitude (pronounced layering-induced anisotropy), the K_r estimate should be within a factor of three of the K_r of the formation. However, as discussed in the previous section, the existence of very pronounced anisotropy should be evident from the results of an analysis with the Cooper et al. method. Thus, uncertainty about the anisotropy ratio should not be the source of large errors in the K_r estimate.

As with the fully penetrating variant of the Hvorslev method, a second issue of practical importance is that of fitting a straight line to the semilog plot of the test data. However, this issue is of less significance in the case of a partially penetrating well because vertical flow suppresses the concave-upward curvature of the response data. Thus, an approximate linear form for the log head vs. time plot will exist at considerably higher α values than in the fully penetrating case. In order to minimize possible errors introduced by the concave-upward curvature, a straight line should be fit to normalized data over the interval of 0.15-0.25 as recommended by Butler (1996).

A third issue of practical importance is that of boundaries in the vertical plane. The mathematical model defined by Equations (5.17a) to (5.17f) is based on the assumption of an infinitely thick formation (Equations [5.17c] and [5.17f]). Obviously, in reality, the formation will be of finite thickness. Hvorslev (1951) presents an approach for the analysis of tests performed in a well in which the screen abuts against an impermeable boundary (case 7 of Hvorslev [1951]). The equation for estimation of K_r in this case can be written as

$$K_r = \frac{r_c^2 \ln\left[1/\psi + \left(1 + (1/\psi)^2\right)^{1/2}\right]}{2bT_0} \tag{5.20}$$

Note that the only difference with Equation (5.19) is that the two has been removed from in front of the ψ parameter. Hyder et al. (1994) have shown that Equation (5.20) is only needed for wells of small aspect ratios (on the order of 10 to 20 or less) screened in isotropic to very slightly anisotropic formations.

A fourth issue of practical importance is that of wells of very small aspect ratios. As discussed by Hvorslev (1951), error may be introduced into the K_r estimates obtained with Equations (5.19) and (5.20) as the aspect ratio approaches one. This error, however, should be less than 15% for open holes or well screens without a bottom cap. Somewhat larger errors may be expected in cases where the well screen has a solid bottom cap. For a standpipe that is only open to the formation at the bottom, Hvorslev presents an empirically derived formula that can be manipulated to obtain the following expression for the estimation of K_r:

$$K_r = \left(\frac{\pi r_c}{5.50\sqrt{K_z T_0}} \right)^2 \tag{5.21}$$

A final issue of practical importance is that of the relative speed of slug introduction. As in the fully penetrating case, the partially penetrating variant of the Hvorslev method does not require that the slug be introduced in a near-instantaneous fashion relative to the formation response. As long as the storage properties of the well-formation configuration are not significantly affecting test responses and/or the hydraulic conductivity of the formation is not extremely large, the relative speed of slug introduction should not be a major concern.

The partially penetrating variant of the Hvorslev method can be illustrated using the Pratt County slug test discussed in the previous section. Figure 5.15A displays a plot of the test data in a log normalized head vs. time format; the near-linear form is a common characteristic of slug tests performed in partially penetrating wells, and is usually an indication of a significant component of vertical flow. Hyder et al. (1994) point out, however, that a near-linear plot can also be an indication of a low-permeability well skin, a topic that will be discussed further in Chapter 9. Figure 5.15B displays a straight line fit to the normalized head range recommended by Butler (1996). Substitution of the T_0 estimate (18.1 sec) obtained from this fit into Equation (5.19) yields a hydraulic conductivity estimate of 16.1 m/d, a value that is approximately 2.6 times less than the estimate found with the Cooper et al. method. Note that the details of the fitting procedure have little impact on the results in this case because of the near-linear nature of the plot. For example, utilization of a fitting procedure that emphasizes the early portions of the data record yields a hydraulic conductivity estimate of 17.6 m/d, a value that differs by less than 10% from the estimate obtained using a normalized head range of 0.15-0.25.

Confined Extensions of the Dagan Method

Dagan (1978) proposes a method for the analysis of slug tests in partially penetrating wells in unconfined formations. Widdowson et al. (1990) and Cole and Zlotnik (1994) outline extensions of the approach to confined formations. These extensions are based on a mathematical model similar to that defined in Equations (5.17a) to (5.17f). The primary difference between these extensions and the Hvorslev model is that the formation is no longer considered to be infinite in the vertical plane. Instead, the formation is assumed to be bounded on the top and bottom by impermeable layers. Thus, Equations (5.17c) and (5.17f) are replaced by the following conditions:

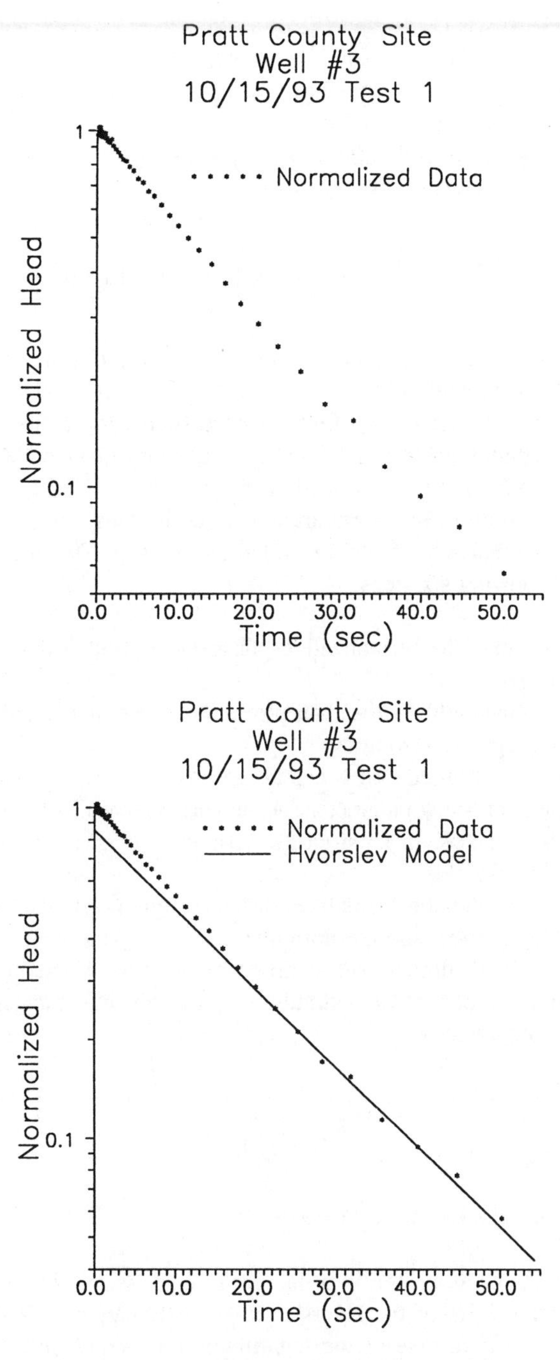

FIGURE 5.15 (A) Logarithm of normalized head ($H(t)/H_0$) vs. time plot of the slug test in well 3 at the Pratt County site; (B) Logarithm of normalized head ($H(t)/H_0$) vs. time plot of the Pratt County test and the Hvorslev model fit for the normalized head range recommended by Butler (1996).

$$\frac{\partial h(r,0,t)}{\partial z} = \frac{\partial h(r,B,t)}{\partial z} = 0, r_w < r < \infty, t > 0 \tag{5.22a}$$

$$\frac{\partial h(\infty,z,t)}{\partial r} = 0, 0 \le z \le B, t > 0 \tag{5.22b}$$

$$\frac{\partial h(r_w,z,t)}{\partial r} = 0, 0 < z < d, d+b < z < B, t > 0 \tag{5.22c}$$

Note that the coordinates in the z direction are with respect to the closest impermeable boundary in this notation.

Widdowson et al. (1990) use a finite-element model to obtain an approximate solution to the mathematical model defined by Equations (5.17a) and (5.17b), (5.17d) and (5.17e), and (5.22a) to (5.22c), while Cole and Zlotnik (1994) obtain a general semianalytical solution to this configuration. In both cases, the solutions result in an analysis method that is very similar to that of Hvorslev. This method essentially consists of the following six steps:

1. The logarithm of the normalized response data is plotted vs. the time since the test began;
2. A straight line is fit to the data plot either via visual inspection or an automated regression routine;
3. The slope of the fitted line is calculated;
4. An estimate of the ψ parameter for the particular well-formation configuration is obtained. In most cases, a value of one is employed for the anisotropy ratio;
5. Given the ψ value, the normalized distance from the closest impermeable boundary ((d+b)/b), and the normalized length of the well screen (b/B), a value for P, the dimensionless flow parameter, is selected;
6. The radial component of hydraulic conductivity is estimated using the following equation:

$$K_r = \frac{r_c^2(1/P)}{2bT_0} \tag{5.23}$$

where: P = dimensionless flow parameter.

Equation (5.23) has been written in a manner to emphasize its similarity with the partially penetrating form of the Hvorslev method (Equation [5.19]). In this case, the P parameter is the inverse of the logarithmic term employed in the Hvorslev method. Widdowson et al. (1990) present tables and figures for the estimation of P. Table 5.5 is similar in form to those presented by Widdowson et al. (1990), generated using the semianalytical solution of Cole and Zlotnik (Cole, pers. commun., 1996).

TABLE 5.5
Tabulated Values of the Dimensionless Flow Parameter, P, Used in Confined Form of the Dagan Method

	(d+b)/b				
ψ	**8.0**	**4.0**	**2.0**	**1.5**	**1.05**
0.20	0.741	0.727	0.681	0.640	0.561
0.10	0.539	0.533	0.505	0.483	0.432
0.067	0.458	0.455	0.432	0.416	0.377
0.050	0.412	0.408	0.390	0.378	0.345
0.033	0.359	0.357	0.343	0.331	0.307
0.025	0.328	0.325	0.314	0.305	0.285
0.020	0.307	0.305	0.295	0.288	0.270
0.013	0.275	0.273	0.263	0.259	0.245
0.010	0.254	0.254	0.246	0.240	0.230
0.0067	0.232	0.230	0.224	0.218	0.211
0.0050	0.218	0.216	0.210	0.205	0.199

Note: Values generated with the semianalytical solution of Cole and Zlotnik [1994], courtesy of K. D. Cole. Values for b/B ≤0.05.

Note that the tabular values essentially do not change for a (d+b)/b ratio greater than 8; so, a normalized distance greater than 8 is considered equivalent to the case of a formation of infinite vertical extent.

Table 5.5 reveals the similarity between the semianalytical solution of Cole and Zlotnik (1994) and the approximate solution of Dachler (1936) upon which the Hvorslev method is based. Note that the P values for a normalized distance of 8 and the inverse of the logarithmic term in Equation (5.19), the Hvorslev equivalent to P, are within 20% of one another. Similar agreement is also seen for the case of a well screened near an impermeable boundary (normalized distance approaching one) and the inverse of the logarithmic term in Equation (5.20). Closer scrutiny of Table 5.5 reveals that the inverse of the logarithmic term in Equation (5.19) actually provides a very reasonable estimate of P, regardless of the distance to the closest impermeable boundary. Thus, in most cases, the relatively small improvement in the K_r estimate that can be obtained from use of the Dagan method is not worth the additional effort required for interpolation of values from Table 5.5.

The Pratt County slug test can be used to illustrate the confined extension of the Dagan method. In this case, the well screen is a normalized distance of approximately 7 from the closest impermeable boundary and the normalized screen length the formation is much less than 0.05. The aspect ratio is approximately 12, producing a ψ value of 0.082 if the formation is assumed to be nearly isotropic. Interpolating from the values of Table 5.5 produces a value of 0.50 for P. Substitution of this value into Equation (5.23) and using the same quantities as employed in the Hvorslev example for the remaining parameters results in a K_r estimate of 12.9 m/d, a value within 20% of the Hvorslev estimate for this same test. Thus, there appears to be reasonable agreement between the conductivity estimates obtained using the Hvorslev and Dagan methods for this test.

The same practical issues discussed with respect to the Hvorslev method are of importance for this approach. Uncertainty about anisotropy, which interval of the response data should be fit with a straight line, the impact of noninstantaneous slug introduction, etc. are all issues of concern for field applications of the Dagan method. Thus, the discussion of those issues presented in the preceding section is equally valid for this method as well.

The KGS Model

Although use of the Cooper et al. method for the analysis of slug tests performed in partially penetrating wells allows the relative importance of the vertical component of flow to be assessed, the parameter estimates obtained with this approach are, in general, of rather limited use. One feature of the Cooper et al. method that separates it from the Hvorslev and Dagan techniques is that the impact of the storage properties of the formation configuration are incorporated into the analysis. In order to retain that feature but also provide a more realistic representation of the flow induced by a slug test in a partially penetrating well, the Cooper et al. method has been extended to the case of a partially penetrating well. The original version of this extension was based on the semianalytical solution of Dougherty and Babu (1984). A more general version of this solution, incorporating anisotropy and both confined and unconfined conditions, was presented by Hyder et al. (1994). This more general solution is designated as the KGS model in this book.

The KGS model for confined formations is based on the mathematical model defined as follows:

$$\frac{\partial^2 h}{\partial r^2} + \frac{1}{r}\frac{\partial h}{\partial r} + \frac{K_z}{K_r}\frac{\partial^2 h}{\partial z^2} = \frac{S_s}{K_r}\frac{\partial h}{\partial t} \quad (5.24a)$$

$$h(r,z,0) = 0, r_w < r < \infty, 0 \le z \le B \quad (5.24b)$$

$$H(0) = H_0 \quad (5.24c)$$

$$h(\infty,z,t) = 0, t > 0, 0 \le z \le B \quad (5.24d)$$

$$\frac{\partial h(r,0,t)}{\partial z} = \frac{\partial h(r,B,t)}{\partial z} = 0, r_w < r < \infty, t > 0 \quad (5.24e)$$

$$\frac{1}{b}\int_{d}^{d+b} h(r_w,z,t)dz = H(t), t > 0 \quad (5.24f)$$

$$2\pi r_w K_r b \frac{\partial h(r_w, z, t)}{\partial r} = \pi r_c^2 \frac{dH(t)}{dt} \square(z),\ t > 0 \tag{5.24g}$$

where: $\square(z)$ = boxcar function = 0, $z < d$, $z > b+d$,
= 1, elsewhere.

Note that other than allowing for a partially penetrating well and the possibility of a vertical component of flow, this model is based on the same set of assumptions that is employed in the mathematical model underlying the Cooper et al. method.

As shown by Hyder et al. (1994), the analytical solution to the mathematical model defined in Equations (5.24a) to (5.24g) can be written as:

$$\frac{H(t)}{H_0} = f(\beta, \alpha, \psi, d/b, b/B) \tag{5.25}$$

This solution, when plotted as normalized head vs. the logarithm of β for a particular well-formation configuration (ψ, d/b, b/B), forms a series of type curves, with each type curve corresponding to a different value of α (Figures 5.16A to 5.16C). The KGS method involves fitting one of the α curves to the field data via manual curve matching or, more commonly, an automated analog. The method consists of the following six steps:

1. The normalized response data are plotted vs. the logarithm of the time since the test began;
2. Estimates of ψ, d/b, and b/B are obtained for the specific well-formation configuration being tested. Usually, as a first guess, the anisotropy ratio incorporated in the ψ parameter is assumed equal to one;
3. As in the Cooper et al. method, the data are overlain by a type-curve plot and the type curves are shifted along the x axis of the data plot until one of the α curves approximately matches the plot of the field data (e.g., Figure 5.3B);
4. Match points are selected from each plot. For convenience's sake, β is set to 1.0 and the real time ($t_{1.0}$) corresponding to $\beta = 1.0$ is read from the x axis of the data plot. An α estimate (α_{cal}) is obtained from the type curve most closely matching the data plot;
5. An estimate of the radial component of hydraulic conductivity is calculated from the definition of β for a partially penetrating well:

$$K_r = \frac{r_c^2}{bt_{1.0}} \tag{5.26a}$$

6. An estimate of the specific storage is calculated from the definition of α for a partially penetrating well:

$$S_s = \frac{\alpha_{cal} r_c^2}{r_w^2 b} \tag{5.26b}$$

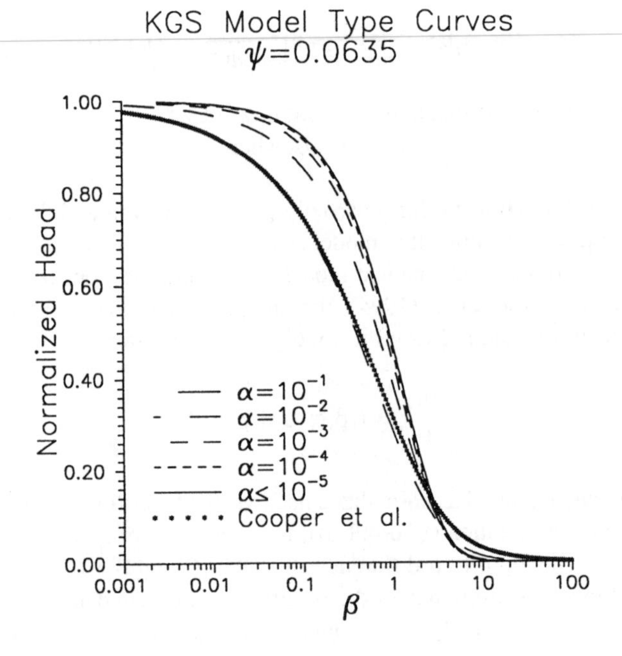

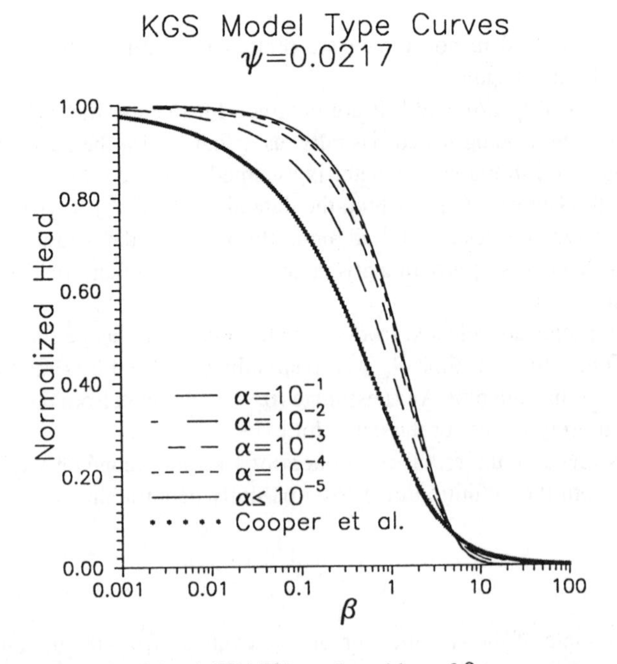

FIGURE 5.16 Normalized head ($H(t)/H_0$) vs. logarithm of β type curves generated using the KGS model for different values of ψ.

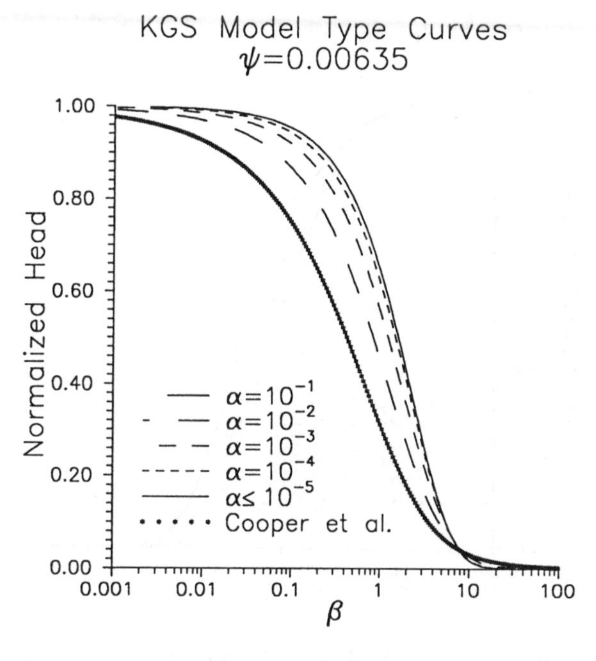

FIGURE 5.16 (continued)

There are two issues of practical importance with respect to the KGS model. First, as with tests in fully penetrating wells, it may be very difficult to obtain a reliable estimate of specific storage from slug tests in partially penetrating wells. In fact, estimation of specific storage to any degree of reliability may be virtually impossible from tests performed in wells with small to moderate aspect ratios (large to moderate ψ). Figures 5.16A-5.16C illustrate how the form of the type curves changes with ψ. In all three figures, type curves for moderate to small values of α can essentially be represented by a single curve, which is crossed by the high α curves at large values of β. This coalescing of type curves at small to moderate values of α is an indication that the elastic storage mechanisms encapsulated in the specific storage parameter are of little importance in this configuration. However, at very large α values, the type curves do converge on those predicted by the Cooper et al. model (the * plot in Figures 5.16A to 5.16C represents the $\alpha = 0.1$ type curve for the Cooper et al. method). Thus, only under a limited range of conditions can a reasonable estimate of specific storage be obtained from a slug test in a partially penetrating well.

Although the solution given by Equation (5.25) does indicate that the slug-induced responses are affected by the anisotropy ratio, it is virtually impossible to estimate the anisotropy ratio in field applications. Figure 5.17 can be used to illustrate the difficulty of estimating the anisotropy ratio from a slug test. This figure depicts the theoretical dependence of normalized responses on the anisotropy ratio. The solid line ($\psi = 0.0635$) displays the normalized responses that would be observed

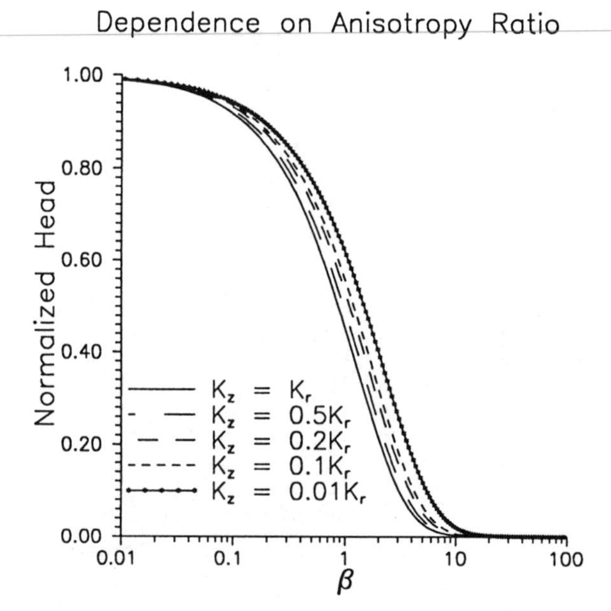

FIGURE 5.17 Normalized head ($H(t)/H_0$) vs. logarithm of β plots illustrating the dependence on the anisotropy ratio (K_z is the vertical component of hydraulic conductivity).

for a well with an aspect ratio of approximately 16 in an isotropic formation, while the rightmost line is for a test performed in a similar configuration except that the vertical component of hydraulic conductivity is two orders of magnitude less than the radial component ($\psi = 0.00635$). Note that the primary effect of the smaller anisotropy ratio is to translate the curve to a larger time; the change in the shape of the curve is quite small. However, this translation could also be produced in an isotropic formation by a smaller value of K_r; so, it is virtually impossible to separate the effect of a decrease in the anisotropy ratio from a decrease in the radial component of hydraulic conductivity. The subtle changes in the shape of the plot of the response data produced by the presence of significant anisotropy will be difficult to exploit given the background noise inherent in field applications. Thus, there is little information about anisotropy that can be gleaned from an analysis using the KGS model. Although some information about the relative importance of the vertical component of flow can be obtained with the Cooper et al. method, it is virtually impossible to quantify the degree of vertical anisotropy using response data from a single well. Use of multiwell slug tests for the estimation of the anisotropy ratio will be discussed in Chapter 10.

The same Pratt County slug test used in the previous sections can be employed to illustrate a field application of the KGS model. Figure 5.18 displays the close match between the data plot and the $\alpha \leq 2.e\text{-}5$ type curve for the case of an isotropic formation. Substituting the $t_{1.0}$ value found from the type curve match into Equation (5.26a) yields an estimate of the radial component of hydraulic conductivity equal

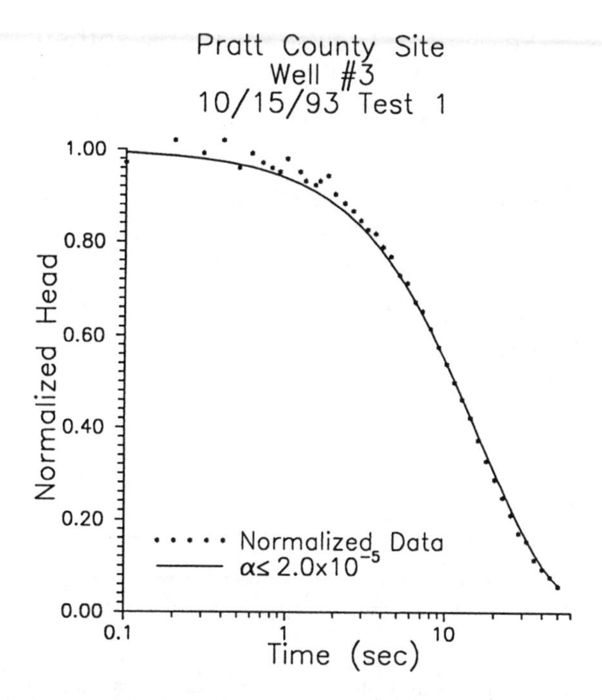

FIGURE 5.18 Normalized head ($H(t)/H_0$) vs. logarithm of time plot of the slug test in well 3 at the Pratt County site and the best-fit KGS Model type curve.

to 17.6 m/d, a value that is approximately a factor of 2.4 less than the K_r estimate obtained with the Cooper et al. method (42.0 m/d). Note that the K_r estimate obtained using the KGS model is within 10% of the value obtained with the Hvorslev method. Substituting ≤2.e-5 for α_{cal} in Equation (5.26b) yields an estimate of specific storage of ≤3.4e-6 m^{-1}, a range that is reasonable for the materials in which the well is screened.

The Peres et al. Approximate Deconvolution Method

The approximate deconvolution method is applied to slug tests performed in partially penetrating wells in the same manner as for tests in fully penetrating wells: the response data are transformed into the equivalent drawdown using Equation (5.13b), the equivalent drawdown data are plotted vs. the logarithm of time, a straight line is fit to the data at moderate to large times, and the Cooper-Jacob semilog method is employed to estimate the hydraulic conductivity of the formation with Equation (5.14b). Although the mechanics of the procedure are the same, there are some important differences between the implementation of this method in fully and partially penetrating wells. These differences can be best demonstrated with a hypothetical example.

Figure 5.19 is an equivalent drawdown plot for a hypothetical slug test performed in a partially penetrating well in the same formation as that used in the earlier

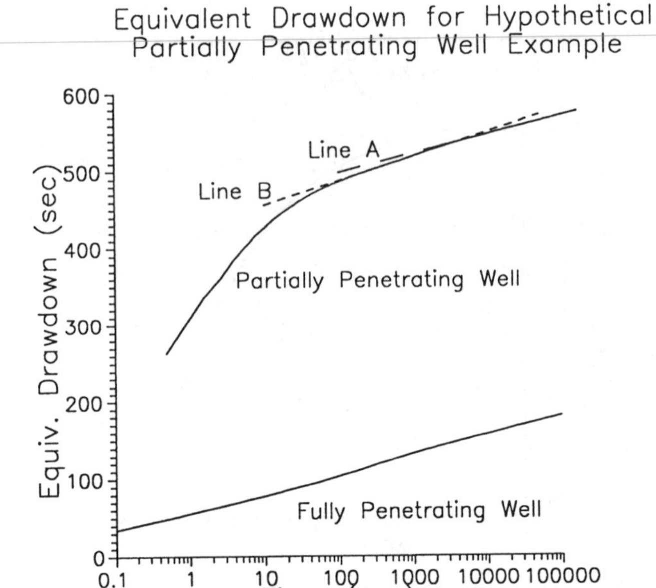

FIGURE 5.19 Equivalent drawdown plot for simulated slug test in same hypothetical aquifer as in Figure 5.9 ($\psi = 0.0635$; equivalent drawdown calculated with the approximate deconvolution method).

example of Figures 5.9A-5.9C. The responses for this hypothetical test were generated using the KGS model with a ψ value of 0.0635. Note that the equivalent drawdown for the case of a fully penetrating well (Figure 5.9C) is provided for comparison.

The most obvious feature of the equivalent drawdown plot for a test in a partially penetrating well is the pronounced concave-downward curvature produced by the additional head losses incurred by the vertical component of the slug-induced flow. This curvature can lead to a systematic underestimation of the hydraulic conductivity of the formation, the magnitude of which will depend on the particulars of the well-formation configuration. Note that the equivalent drawdowns displayed on Figure 5.19 were calculated assuming that noise-free responses are available. Line A is the best-fit straight line obtained when fitting the latter portion of the equivalent drawdown plot, corresponding to a range of normalized heads (<0.01) not normally available in conventional groundwater investigations. When the slope of this line is substituted into Equation (5.14b), a hydraulic conductivity estimate 4.5% lower than the value used to generate the responses is obtained. If relatively noise-free data are only available until a normalized head of approximately 0.01 (Line B), a more reasonable assumption for environmental applications, the hydraulic conductivity estimate obtained from Equation (5.14b) will underpredict the actual value by 26%. Although the error is higher than in the fully penetrating case as a result of the impact of the vertical component of flow, the conductivity estimate may still be quite reasonable for many applications.

The same practical issues discussed earlier in the section on tests in fully penetrating wells are important in the partially penetrating variant of this method. However, there are two additional issues of particular significance for tests in partially penetrating wells. First, as shown in the hypothetical example, the equivalent drawdown plot will asymptotically converge on the semilog straight line. With a thicker formation and/or a vertical anisotropy ratio less than one, the range of equivalent drawdowns for which a reasonable conductivity estimate can be obtained will occur at later times. In some cases this range will correspond to normalized responses that are too small to be obtained in most slug tests performed for shallow groundwater applications. In the majority of cases, however, estimates within 30-40% of the actual conductivity of the formation should be obtainable if the straight line is fit to the final portion of the equivalent drawdown plot. As was emphasized in the section on fully penetrating wells, the technique works best when a relatively large H_0 is used, high accuracy pressure transducers are available, background noise is quite small, and the test is run to complete recovery.

A second issue of practical significance is that of the estimation of the actual thickness of the interval of the formation through which the slug-induced flow is moving (henceforth designated as the flow interval). In all other methods for the analysis of slug tests in partially penetrating wells, one of the most critical issues is the definition of the effective screen length. However, in the approximate deconvolution method, this quantity is of no importance because the method is used only after the impact of the slug-induced disturbance has spread to its full vertical extent and changes in head at the test well are a function of radial flow through the full thickness of the flow interval. Uncertainty about this thickness, however, can introduce considerable error into the conductivity estimate. This uncertainty is often greatest in wells that are screened for a relatively short length in the upper portions of what is suspected to be a much thicker unit. Butler and Healey (1998) present a field example that demonstrates the error that can be introduced into the conductivity estimate through uncertainty about the thickness of the flow interval. As they emphasize, information from drilling logs and geophysical surveys is invaluable in providing constraints on this thickness.

The same Pratt County slug test used in the previous sections can be employed to demonstrate a major limitation of the approximate deconvolution method. Figure 5.13 is a plot of the normalized response data for the Pratt County test, the head fluctuations at early times are accentuated by use of a relatively small H_0 (0.38 m). In this test, data collection was terminated at a normalized head of 0.057. The combination of early-time head fluctuations, a small H_0 value, and termination at a relatively large normalized head make the approximate deconvolution approach of virtually no use for this test. This is illustrated by the equivalent drawdown plot given in Figure 5.20. Although the response data for the first 2 sec of the test were ignored for the purposes of this plot, little if any information about the conductivity of the formation can be gained. Even the equivalent drawdown corresponding to moderate to low normalized heads is relatively noisy because of the small H_0 coupled with the use of a low-accuracy pressure transducer. Thus, a large H_0 and a high-accuracy transducer are strongly recommended in any tests for which the approximate deconvolution method might be applied.

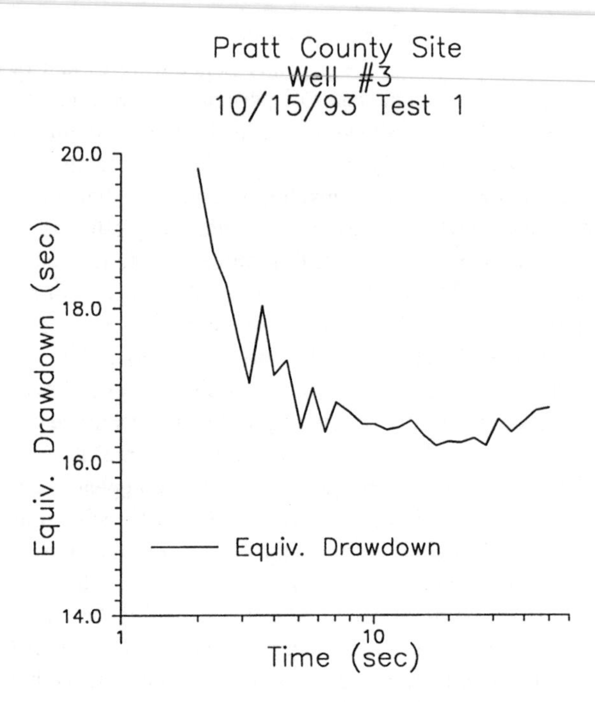

FIGURE 5.20 Equivalent drawdown vs. log time plot of the slug test in well 3 at the Pratt County site (equivalent drawdown calculated with the approximate deconvolution method).

Additional Methods

The majority of slug tests performed in partially penetrating wells in confined formations are analyzed using one of the preceding approaches. However, there are several other techniques that have been employed in the literature for this same purpose. The three most common are the method of Bouwer and Rice (1976), the method of Nguyen and Pinder (1984), and that of Dax (1987). Each of these methods is briefly described in the following paragraphs.

The method of Bouwer and Rice (1976) was developed for the analysis of slug tests in partially penetrating wells in unconfined formations. However, Bouwer (1989) states that the approach can also be employed for confined formations, and applications of the method for this purpose are not infrequent (e.g., Campbell et al., 1990; Brother and Christians, 1993). As long as the test interval is not close to the upper boundary, which is assumed to be the water table in the Bouwer and Rice method, K_r estimates obtained with this approach should be in reasonable agreement with values obtained using the Hvorslev method. The Bouwer and Rice method will be discussed at length in the following chapter on tests in unconfined formations.

The method of Nguyen and Pinder (1984) was one of the first approaches proposed for the analysis of response data from slug tests in partially penetrating wells in confined formations that incorporated the elastic storage mechanisms encapsulated in the specific storage parameter. Butler and Hyder (1994) have shown, however, that the parameter estimates obtained using the Nguyen and Pinder method must be viewed with considerable skepticism owing to an error in the analytical solution to the mathematical model on which the method is based. Thus, the Nguyen and Pinder method is not recommended for the analysis of slug-test data.

The method of Dax (1987) for slug tests in fully penetrating wells has been extended to the case of partially penetrating wells. However, the extension ignores flow in the vertical direction; so, this approach is not recommended for the analysis of response data from partially penetrating wells.

6 The Analysis of Slug Tests — Unconfined Formations

CHAPTER OVERVIEW

The most common application of slug tests at sites of suspected groundwater contamination is in shallow wells screened in unconfined flow systems. The emphasis of this chapter will be on the methods that have been developed for the analysis of response data from slug tests performed in such settings. Specialized techniques developed for the analysis of tests in unconfined formations in which the hydraulic conductivity is very high or very low, or in cases where the response data are impacted by a well skin, will be discussed in later chapters.

SLUG TESTS IN UNCONFINED FORMATIONS

The major methods for the analysis of response data from slug tests in unconfined formations can all be discussed in the context of the hypothetical cross sections illustrated in Figure 6.1. The methods are primarily classified on the basis of whether or not the well screen (or the filter pack) intersects the water table. If the well is screened below the water table (Figure 6.1A), the change in the saturated thickness during a test is usually quite small and the relevant physics can be represented by a linear mathematical model. If the screen intersects the water table (Figure 6.1B), the assumption of a constant saturated thickness or a constant effective screen length may no longer be appropriate and a nonlinear mathematical model may be required to represent the relevant physics. In the case of a well screened below the water table, the methods can be further subdivided on the basis of their conceptualization of the storage mechanism.

WELLS SCREENED BELOW THE WATER TABLE

In hydrogeologic investigations, the vast majority of slug tests performed in unconfined formations in wells screened below the water table are analyzed with one of the following three techniques: the method of Bouwer and Rice (1976), the method of Dagan (1978), and the method based on the unconfined variant of the KGS model (Hyder et al., 1994). Each of these techniques is described in the following sections, after which a less commonly utilized approach is briefly summarized.

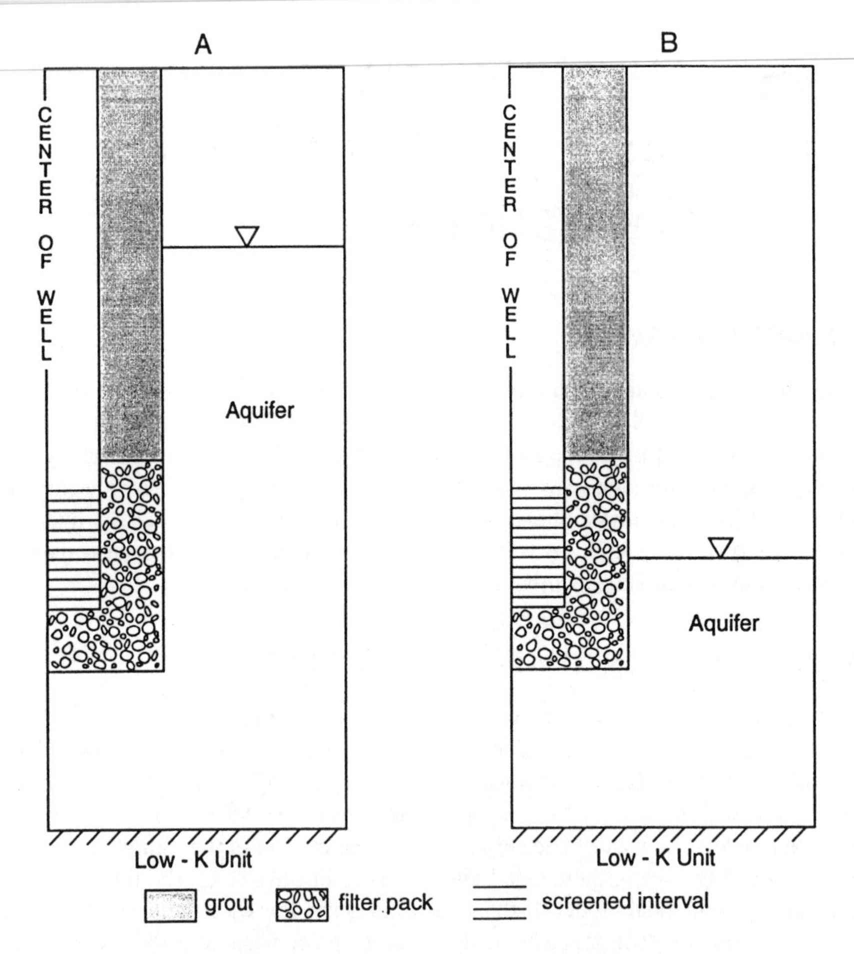

FIGURE 6.1 Hypothetical cross sections depicting a well screened below the water table (A) and a well screened across the water table (B). (Figures not to scale.)

The Bouwer and Rice Method

The Bouwer and Rice method is based on the mathematical model defined as follows:

$$\frac{\partial^2 h}{\partial r^2} + \frac{1}{r}\frac{\partial h}{\partial r} + \frac{K_z}{K_r}\frac{\partial^2 h}{\partial z^2} = 0 \quad (6.1a)$$

$$H(0) = H_0 \quad (6.1b)$$

$$h(r,0,t) = 0,\ r_w < r < R_e, t > 0 \quad (6.1c)$$

$$\frac{\partial h(r,B,t)}{\partial z} = 0,\ r_w < r < R_e,\ t > 0 \quad (6.1d)$$

$$h(R_e, z, t) = 0,\ 0 \le z \le B,\ t > 0 \tag{6.1e}$$

$$h(r_w, z, t) = H(t),\ d \le z \le (d + b),\ t > 0 \tag{6.1f}$$

$$2\pi r_w K_r \int_d^{d+b} \frac{\partial h(r_w, z, t)}{\partial r} dz = \pi r_c^2 \frac{dH(t)}{dt}, t > 0 \tag{6.1g}$$

$$\frac{\partial h(r_w, z, t)}{\partial r} = 0,\ 0 \le z < d,\ d + b < z \le B,\ t > 0 \tag{6.1h}$$

Two key assumptions of this mathematical model are (1) the effects of elastic storage mechanisms can be neglected (Equation 6.1a), and (2) the position of the water table, and thus the saturated thickness of the formation, does not change during the course of a test (Equation 6.1c). Although the original mathematical model of Bouwer and Rice was defined for an isotropic formation, the extension of Zlotnik (1994) to the general anisotropic case is considered here. Note that this model, similar to that upon which the fully penetrating variant of the Hvorslev method is based, employs an empirical parameter (R_e) that is designated as the effective radius of the slug test.

The analytical solution to the mathematical model defined by Equations (6.1a) to (6.1h) can be written as (Bouwer and Rice, 1976; Zlotnik, 1994):

$$\ln\left(\frac{H(t)}{H_0}\right) = -\frac{2K_r bt}{r_c^2 \ln(R_e/r_w^*)} \tag{6.2}$$

where: $r_w^* = r_w (K_z/K_r)^{1/2}$.

As with the solutions that are the basis of the methods of Hvorslev and Dagan, an important feature of Equation (6.2) is that a plot of the logarithm of normalized head vs. time is a straight line. Thus, the Bouwer and Rice method also involves calculating the slope of a straight line fit to the response data and using that value to estimate the hydraulic conductivity of the formation. The method essentially consists of the following five steps:

1. The logarithm of the normalized response data is plotted vs. the time since the test began;
2. A straight line is fit to the data plot either via visual inspection or an automated regression routine;
3. The slope of the fitted line is calculated. If the time lag (T_0) is used in this calculation, the slope, when written in terms of the natural logarithm, becomes $-1/T_0$;

4. Values for the anisotropy ratio and the effective radius parameter are estimated for the particular well-formation configuration. Unless information exists to the contrary, the anisotropy ratio is assumed equal to one;
5. The radial component of hydraulic conductivity is estimated with an expression obtained by rearranging Equation (6.2) and rewriting it in terms of the slope calculated using T_0:

$$K_r = \frac{r_c^2 \ln\left[R_e/r_w^*\right]}{2bT_0} \tag{6.3}$$

Note the similarity between Equation (6.3) and Equations (5.7), (5.19), and (5.23). These expressions differ only in the factor appearing in the numerator on the right-hand side. This similarity is not unexpected given that all of these expressions are based on a quasi-steady state representation of the slug-induced flow.

There are several issues of practical importance with respect to the Bouwer and Rice method. The issue of most significance is that of the estimation of R_e. As with the fully penetrating variant of the Hvorslev method, R_e should be viewed as an empirical parameter, and not as the actual effective radius of a slug test. Bouwer and Rice (1976) present expressions for the estimation of R_e that are based on a series of electrical-analog simulations of the mathematical model defined in Equations (6.1a) to (6.1h). These expressions are written in terms of $\ln(R_e/r_w^*)$, the quantity that appears in the numerator of Equation (6.3), and not R_e alone. For a well that terminates above the lower impermeable boundary (a "partially penetrating" well in the terminology of their paper), Bouwer and Rice present the following expression for estimation of $\ln(R_e/r_w^*)$:

$$\ln\left(R_e/r_w^*\right) = \left[\frac{1.1}{\ln\left((d+b)/r_w^*\right)} + \frac{\boldsymbol{A} + \boldsymbol{B}\left(\ln\left[\left(B-(d+b)\right)/r_w^*\right]\right)}{b/r_w^*}\right]^{-1} \tag{6.4a}$$

where: $\boldsymbol{A},\boldsymbol{B}$ = empirical coefficients, [dimensionless].

Bouwer and Rice note that when the $\ln[(B - (d + b))/r_w^*]$ term is greater than 6.0, 6.0 should be used instead of the actual value. For a well that terminates at the lower impermeable boundary (a "fully penetrating" well in the terminology of their paper), Bouwer and Rice present the following expression:

$$\ln\left(R_e/r_w^*\right) = \left[\frac{1.1}{\ln\left((d+b)/r_w^*\right)} + \frac{\boldsymbol{C}}{b/r_w^*}\right]^{-1} \tag{6.4b}$$

where: $\boldsymbol{C}$ = empirical coefficient, [dimensionless].

The empirical coefficients appearing in Equations (6.4a) and (6.4b) can be estimated from the modified aspect ratio (b/r_w^*), which is the inverse of ψ, using a figure that is presented in Bouwer and Rice (1976). As reported by Boak (1991), Van Rooy (1988) fit polynomial functions to the curves shown in that figure to develop the following expressions for the empirical coefficients:

$$\begin{aligned} A = {} & 1.4720 + 3.537\times10^{-2}\left(b/r_w^*\right) - 8.148\times10^{-5}\left(b/r_w^*\right)^2 \\ & + 1.028\times10^{-7}\left(b/r_w^*\right)^3 - 6.484\times10^{-11}\left(b/r_w^*\right)^4 \\ & + 1.573\times10^{-14}\left(b/r_w^*\right)^5 \end{aligned} \tag{6.5a}$$

$$\begin{aligned} B = {} & 0.2372 + 5.151\times10^{-3}\left(b/r_w^*\right) - 2.682\times10^{-6}\left(b/r_w^*\right)^2 \\ & - 3.491\times10^{-10}\left(b/r_w^*\right)^3 + 4.738\times10^{-13}\left(b/r_w^*\right)^4 \end{aligned} \tag{6.5b}$$

$$\begin{aligned} C = {} & 0.7920 + 3.993\times10^{-2}\left(b/r_w^*\right) - 5.743\times10^{-5}\left(b/r_w^*\right)^2 \\ & + 3.858\times10^{-8}\left(b/r_w^*\right)^3 - 9.659\times10^{-12}\left(b/r_w^*\right)^4 \end{aligned} \tag{6.5c}$$

Figure 6.2 is a plot of these expressions for aspect ratios from 4 to 1000. It is important to emphasize that regardless of whether Equations (6.5a) to (6.5c) or the original figure given in Bouwer and Rice (1976) are used to estimate the coefficients of Equations (6.4a) and (6.4b), R_e is an empirical parameter; so, relatively little physical significance should be attached to its magnitude.

Many of the same practical issues discussed with respect to the Hvorslev method are of importance for the Bouwer and Rice method. These issues include that of uncertainty about anisotropy, which interval of the response data to fit with a straight line, and the effect of noninstantaneous slug introduction. Hyder and Butler (1995) assess the impact of several key assumptions underlying the mathematical model on which the Bouwer and Rice method is based. They found that for wells of small aspect ratios the assumption of isotropy can result in an underestimation of hydraulic conductivity on the order of a factor of two or more in highly anisotropic systems. However, except in the case of interbedded high- and low-conductivity material, the anisotropy ratio for natural systems should lie between 0.3 and 1.0 (Freeze and Cherry, 1979). Given that range, errors in the K_r estimate introduced through uncertainty about anisotropy should not exceed 20%. An idea of the significance of anisotropy at any particular well can be obtained by a preliminary analysis of the response data with the Cooper et al. model. If there is a high degree of vertical anisotropy, the test data should be closely matched by a type curve for a physically plausible α. Note that if an estimate of the anisotropy ratio is available from core data or other sources, that estimate can be incorporated into the Bouwer and Rice method using Equation (6.3) as shown by Zlotnik (1994).

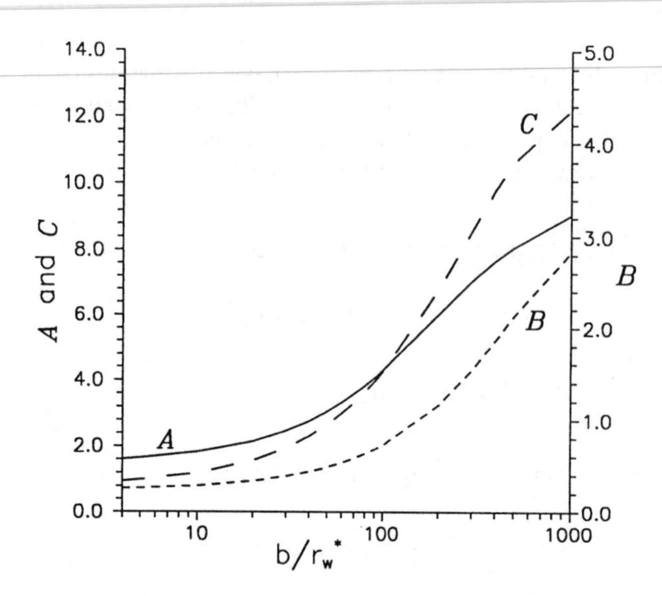

FIGURE 6.2 Plot of empirical coefficients for Bouwer and Rice method generated using the regression equations of Van Rooy (1988).

Hyder and Butler (1995) present results that indicate that the Bouwer and Rice method will significantly overestimate K_r when α is large (Figure 6.3). Their findings, however, were obtained using a fitting scheme that emphasized the early portions of the data record, i.e., the scheme illustrated in Figure 5.8A. If the scheme illustrated in Figure 5.8B is employed, the Bouwer and Rice method will produce reasonable estimates for virtually all values of α. Butler (1996) recommends that normalized heads in the range of 0.20 to 0.30 be used to obtain the best results with the Bouwer and Rice method.

As with the techniques of Hvorslev and Dagan, the Bouwer and Rice method does not strictly require that the slug be introduced in a manner that can be considered near-instantaneous relative to the formation response. Noninstantaneous slug introduction, however, can introduce error into hydraulic conductivity estimates when α is large. Thus, as a general rule, every effort should be made to ensure that the slug is introduced in as rapid a manner as possible.

A field example can be used to illustrate the Bouwer and Rice method. In October of 1993, a series of slug tests were performed at a monitoring well in Pratt County, Kansas (Butler et al., 1993). The well was screened in an unconsolidated alluvial sequence consisting primarily of sands and gravels with interbedded clays. Table 6.1 summarizes the well-construction information, while Table 6.2 lists the test data employed in the analysis. Figure 6.4A displays a near-linear plot of the response data in a log normalized head vs. time format. Figure 6.4B shows a straight line fit to the normalized head range recommended by Butler (1996). The T_0 estimate obtained from this straight line is 66.3 s. Since the well is screened a considerable distance above the lower impermeable boundary, Equation (6.4a) is the appropriate expression for calculation of the $\ln(R_e/r_w^*)$ term. Given the assumption of an isotropic

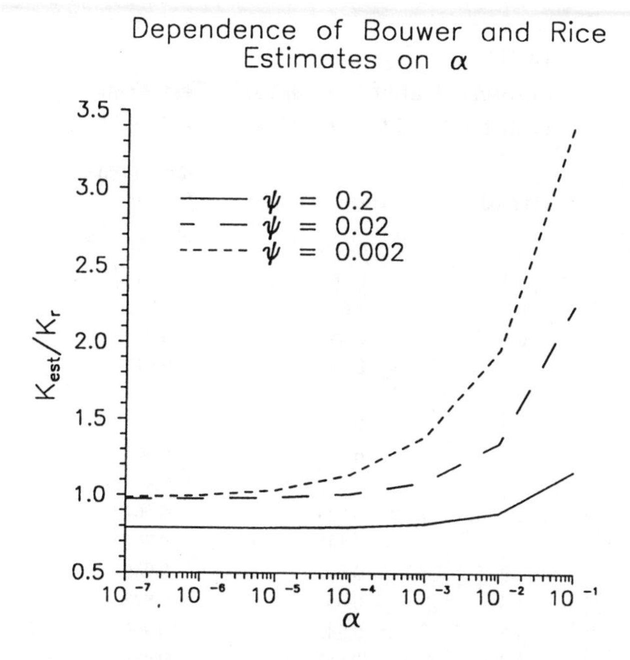

FIGURE 6.3 Plot of hydraulic conductivity ratio (Bouwer and Rice estimate (K_{est}) over actual conductivity (K_r)) vs. logarithm of α as a function of ψ for the case of a partially penetrating well screened at the center of a very thick formation. (After Hyder and Butler, 1995.)

TABLE 6.1
Well Construction Information for Well 4 at Pratt County Monitoring Site 36

Well Designation	r_w(m)	r_c(m)	b(m)	B(m)	d(m)
Pratt County Site 36 Well 4	0.125	0.064	1.52	47.87	16.77

formation and the well-construction parameters of Table 6.1, values of 5.5, 1.96, and 0.267 are calculated for $\ln[(B - (d + b))/r_w^*]$, $\boldsymbol{A}$, and $\boldsymbol{B}$, respectively. Using these values and the well-construction parameters of Table 6.1 in Equation (6.4a) produces a $\ln(R_e/r_w^*)$ value of 2.01. Substitution of the $\ln(R_e/r_w^*)$ and T_0 values into Equation (6.3) yields a K_r estimate of 3.46 m/d. As in the partially penetrating well example of the previous chapter, the details of the fitting procedure have a very limited impact on the K_r estimate because of the near-linear form of the data plot. For example, utilization of a fitting procedure that emphasizes the early portions of the data record yields a hydraulic conductivity estimate of 3.78 m/d, a value that differs by less than

TABLE 6.2
Response Data from 10/15/93 Test #2 in Well 4 at Pratt County Monitoring Site 36

Time (s)	Head (m)	Normalized Head
0.1	0.663	0.988
0.2	0.664	0.989
0.3	0.656	0.978
0.4	0.656	0.978
0.5	0.656	0.978
0.6	0.656	0.977
0.7	0.653	0.973
0.8	0.649	0.967
0.9	0.649	0.967
1.1	0.649	0.967
1.2	0.645	0.962
1.3	0.642	0.956
1.5	0.653	0.973
1.6	0.648	0.966
1.8	0.642	0.956
2.0	0.638	0.951
2.3	0.630	0.940
2.6	0.627	0.934
2.9	0.619	0.923
3.2	0.619	0.923
3.6	0.616	0.918
4.0	0.608	0.907
4.5	0.605	0.902
5.1	0.596	0.889
5.7	0.590	0.879
6.4	0.587	0.874
7.1	0.579	0.863
8.0	0.568	0.846
9.0	0.560	0.835
10.0	0.553	0.824
11.3	0.539	0.803
12.6	0.531	0.791
14.2	0.517	0.770
15.9	0.501	0.747
17.8	0.486	0.724
20.0	0.469	0.698
22.4	0.450	0.671
25.2	0.435	0.648
28.2	0.413	0.615
31.7	0.390	0.581
35.5	0.368	0.549
39.9	0.346	0.516

TABLE 6.2 (continued)
Response Data from 10/15/93 Test #2 in Well 4 at Pratt County Monitoring Site 36

Time (s)	Head (m)	Normalized Head
44.7	0.321	0.479
50.2	0.295	0.440
56.3	0.273	0.407
63.1	0.244	0.363
70.8	0.221	0.329
79.5	0.191	0.285
89.2	0.166	0.247
100.1	0.140	0.208
112.3	0.118	0.176
125.9	0.099	0.148
141.3	0.081	0.120
158.5	0.059	0.088
177.9	0.051	0.077
199.6	0.037	0.055
223.9	0.025	0.037
251.2	0.019	0.028
281.9	0.014	0.021
316.3	0.008	0.011
354.9	0.008	0.011

10% from the estimate obtained with the normalized head range recommended by Butler (1996).

The Dagan Method

The Dagan method is based on a mathematical model similar to that defined in Equations (6.1a) to (6.1h). The only difference is that the model employed in the Dagan method does not assume a constant-head boundary at a finite radial distance from the test well. Instead, hydrologic boundaries in the lateral plane are assumed to be at an infinite distance from the well. This difference is manifested by rewriting Equations (6.1c) to (6.1e) as:

$$h(r,0,t)=0,\ r_w<r<\infty,\ t>0 \tag{6.1cc}$$

$$\frac{\partial h(r,B,t)}{\partial z}=0,\ r_w<r<\infty,\ t>0 \tag{6.1dd}$$

$$\frac{\partial h(\infty,z,t)}{\partial r}=0,\ 0\le z\le B,\ t>0 \tag{6.1ee}$$

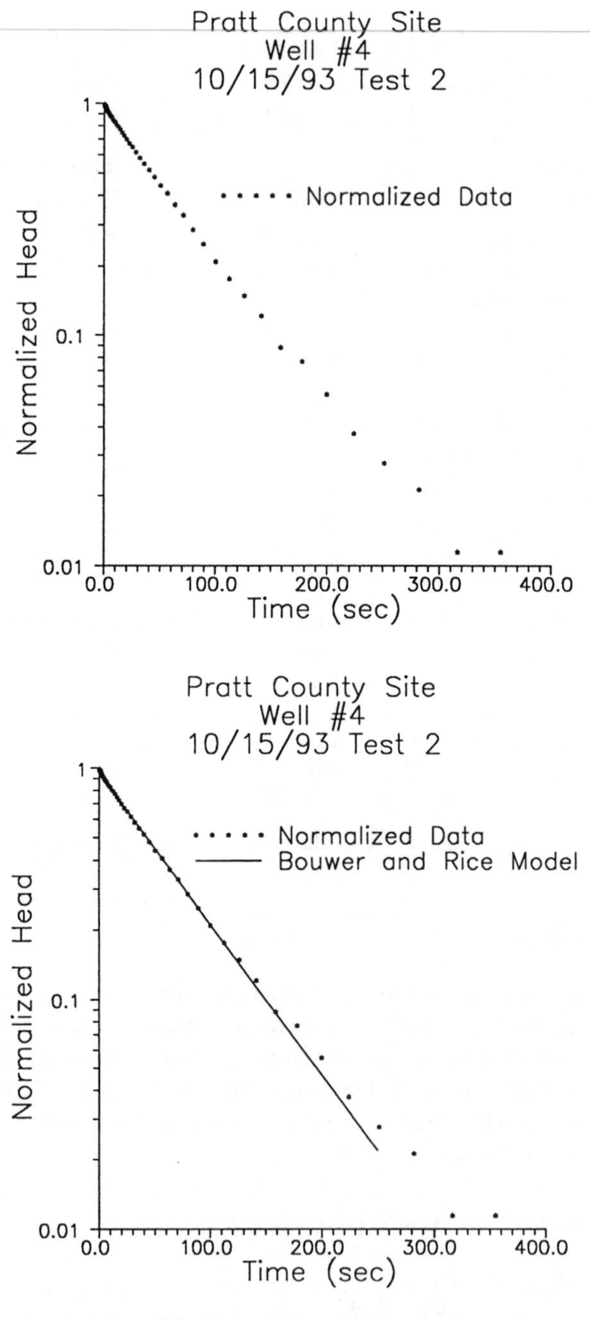

FIGURE 6.4 (A) Logarithm of normalized head ($H(t)/H_0$, where $H(t)$ is deviation from static and H_0 is magnitude of the initial displacement) vs. time plot of a slug test performed in well 4 at a Pratt County, Kansas monitoring site; (B) Logarithm of normalized head ($H(t)/H_0$) vs. time plot of the Pratt County test and the Bouwer and Rice model fit for the normalized head range recommended by Butler (1996).

In this section, all discussions of the mathematical model defined by Equations (6.1a) to (6.1h) assume that Equations (6.1c) to (6.1e) have been replaced by Equations (6.1cc) to (6.1ee).

Dagan (1978) proposes an approximate semianalytical solution to the mathematical model defined by Equations (6.1a) to (6.1h) for aspect ratios greater than 50. Widdowson et al. (1990) use a finite element model to extend the approach to aspect ratios less than 50. Cole and Zlotnik (1994) present a general semianalytical solution for any aspect ratios. In all cases, the analysis approach is similar to that employed in the confined variant of the Dagan method. This approach essentially consists of the following six steps:

1. The logarithm of the normalized response data is plotted vs. the time since the test began;
2. A straight line is fit to the data plot either via visual inspection or an automated regression routine;
3. The slope of the fitted line is calculated;
4. An estimate of the ψ parameter for the particular well-formation configuration is obtained. In most cases, a value of one is used for the anisotropy ratio;
5. Given the ψ value, the normalized distance from the water table ((d+b)/b), and the normalized length of the well screen (b/B), a value for P, the dimensionless flow parameter, is selected;
6. The radial component of hydraulic conductivity is estimated using the following equation:

$$K_r = \frac{r_c^2(1/P)}{2bT_0} \tag{6.6}$$

where: P = dimensionless flow parameter.

Widdowson et al. (1990) present tables and a figure for the calculation of P. Table 6.3 is similar in form to those presented by Widdowson et al. (1990), generated using the semianalytical solution of Cole and Zlotnik (Cole, pers. commun., 1996). Note that the P values essentially do not change for (d+b)/b greater than eight; so, a normalized distance of greater than eight is considered equivalent to the case of a well screen at an infinite distance from the water table. Note also that these results pertain to a well with a normalized screen length (b/B) that is less than or equal to 0.05. Table 6.4 shows how P depends on the normalized screen length for the case of a screen abutting against a lower impermeable boundary.

Many of the practical issues discussed with respect to the Bouwer and Rice method are of importance for this approach as well. Uncertainty about anisotropy, which interval to fit with a straight line, and the impact of noninstantaneous slug introduction are again the primary issues of concern. Thus, the discussion of these issues presented for the Bouwer and Rice method is equally valid for this technique as well.

TABLE 6.3
Tabulated Values of the Dimensionless Flow Parameter, P, Used in Dagan Method for Wells Screened Below the Water Table

	(d+b)/b				
ψ	**8.0**	**4.0**	**2.0**	**1.5**	**1.05**
0.20	0.646	0.663	0.705	0.756	1.045
0.10	0.477	0.487	0.505	0.531	0.687
0.067	0.409	0.416	0.429	0.446	0.562
0.050	0.367	0.373	0.385	0.397	0.491
0.033	0.322	0.325	0.335	0.352	0.414
0.025	0.294	0.297	0.305	0.322	0.370
0.020	0.276	0.278	0.287	0.301	0.342
0.013	0.247	0.249	0.255	0.269	0.300
0.010	0.230	0.231	0.238	0.250	0.276
0.0067	0.211	0.210	0.213	0.227	0.248
0.0050	0.198	0.199	0.201	0.213	0.230

Note: Values generated with the semianalytical solution of Cole and Zlotnik (1994), courtesy of K. D. Cole. Values for b/B ≤0.05.

TABLE 6.4
Tabulated Values of the Dimensionless Flow Parameter, P, Used in Dagan Method for Wells Screened Below the Water Table

	b/B					
ψ	**1.0**	**0.83**	**0.67**	**0.50**	**0.20**	**0.10**
0.20	1.289	0.723	0.631	0.576	0.510	0.492
0.10	0.800	0.510	0.460	0.428	0.390	0.380
0.050	0.536	0.384	0.354	0.335	0.312	0.306
0.025	0.388	0.305	0.286	0.273	0.258	0.254
0.010	0.279	0.238	0.227	0.219	0.209	0.206

Note: Values generated with the semianalytical solution of Cole and Zlotnik (1994), courtesy of K. D. Cole. Values for (d+b) = B.

The Pratt County slug test described in the previous section can be used to illustrate a field application of the Dagan method. In this case, the well screen is at a normalized distance of approximately 12 from the water table and the normalized screen length is less than 0.05. The aspect ratio is approximately 12, producing a ψ value of 0.082 if the formation is assumed to be nearly isotropic. Interpolating from the entries of Table 6.3 produces a value of 0.440 for P. Substitution of this value into Equation (6.6) and using 66.3 for T_0 results in a K_r estimate of 3.91 m/d, a value approximately 13% higher than the Bouwer and Rice estimate obtained for the same test. Note that Dagan (1978) found that the K_r estimates obtained using this approach are consistently larger than those obtained with the Bouwer and Rice method, a finding in agreement with the results of this field example.

The KGS Model

The KGS model for unconfined formations is based on the mathematical model defined as follows:

$$\frac{\partial^2 h}{\partial r^2} + \frac{1}{r}\frac{\partial h}{\partial r} + \frac{K_z}{K_r}\frac{\partial^2 h}{\partial z^2} = \frac{S_s}{K_r}\frac{\partial h}{\partial t} \tag{6.7a}$$

$$h(r,z,0) = 0,\ r_w < r < \infty,\ 0 \le z \le B \tag{6.7b}$$

$$H(0) = H_0 \tag{6.7c}$$

$$h(\infty,z,t) = 0,\ t > 0,\ 0 \le z \le B \tag{6.7d}$$

$$h(r,0,t) = 0,\ r_w < r < \infty,\ t > 0 \tag{6.7e}$$

$$\frac{\partial h(r,B,t)}{\partial z} = 0,\ r_w < r < \infty,\ t > 0 \tag{6.7f}$$

$$\frac{1}{b}\int_{d}^{d+b} h(r_w,z,t)\,dz = H(t),\ t > 0 \tag{6.7g}$$

$$2\pi r_w K_r b \frac{\partial h(r_w,z,t)}{\partial r} = \pi r_c^2 \frac{dH(t)}{dt}\,\square(z),\ t > 0 \tag{6.7h}$$

where: $\square(z)$ = boxcar function = 0, z<d, z>b+d,
= 1, elsewhere.

Note that unlike the Bouwer and Rice or Dagan methods, the KGS model incorporates the storage properties of the media into the analysis (Equation 6.7a), and requires that the slug be introduced in a near-instantaneous fashion relative to the formation response (Equations 6.7b and 6.7c).

As shown by Hyder et al. (1994), the analytical solution to the mathematical model defined in Equations (6.7a) to (6.7h) can be written as:

$$\frac{H(t)}{H_0} = f(\beta,\alpha,\psi,d/b,b/B) \tag{6.8}$$

This solution, when plotted as normalized head vs. the logarithm of β for a particular well-formation configuration (ψ, d/b, b/B), forms a series of type curves, with each type curve corresponding to a different value of α (e.g., Figures 5.16A to 5.16C). As in the confined case, the analysis method involves fitting one of the α curves to the field data via manual curve matching or, more commonly, an automated analog. The method consists of the following six steps:

1. The normalized response data are plotted vs. the logarithm of the time since the test began;
2. Estimates for ψ, d/b, and b/B are obtained for the specific well-formation configuration being tested. Unless information exists to the contrary, the anisotropy ratio incorporated in the ψ parameter is assumed equal to one;
3. As in the confined variant of the KGS model and the Cooper et al. method, the data curve is overlain by a type-curve plot and the type curves are shifted along the x axis of the data plot until one of the α curves approximately matches the plot of the field data (e.g., Figure 5.3B);
4. Match points are selected from each plot. For convenience's sake, β is set to 1.0 and the real time ($t_{1.0}$) corresponding to $\beta = 1.0$ is read from the x axis of the data plot. An α estimate (α_{cal}) is obtained from the type curve most closely matching the data plot;
5. An estimate for the radial component of hydraulic conductivity is calculated from the definition of β for a partially penetrating well:

$$K_r = \frac{r_c^2}{bt_{1.0}} \tag{6.9a}$$

6. An estimate for the specific storage is calculated from the definition of α for a partially penetrating well:

$$S_s = \frac{\alpha_{cal} r_c^2}{r_w^2 b} \tag{6.9b}$$

There are two issues of practical importance with respect to the unconfined variant of the KGS model. First, it will be very difficult to obtain a reliable estimate of specific storage with this method, especially from tests in wells with small to moderate aspect ratios. Second, it will be virtually impossible to obtain an estimate of the anisotropy ratio with data from a single-well slug test. Both of these issues are discussed at length in the section on the confined variant of the KGS model in Chapter 5; so, no further discussion is presented here.

The Pratt County slug test discussed in the previous sections can be used to illustrate a field application of the KGS model for the analysis of data from wells in unconfined formations. Figure 6.5 displays the excellent match between the data plot and the $\alpha = 3.1 \times 10^{-3}$ type curve for assumed isotropic conditions. Substituting the $t_{1.0}$ value found from the type curve match into Equation (6.9a) yields an estimate of K_r equal to 4.21 m/d, a value within 8% of the estimate obtained using the Dagan method. Note that the Bouwer and Rice K_r estimate was approximately 18% lower

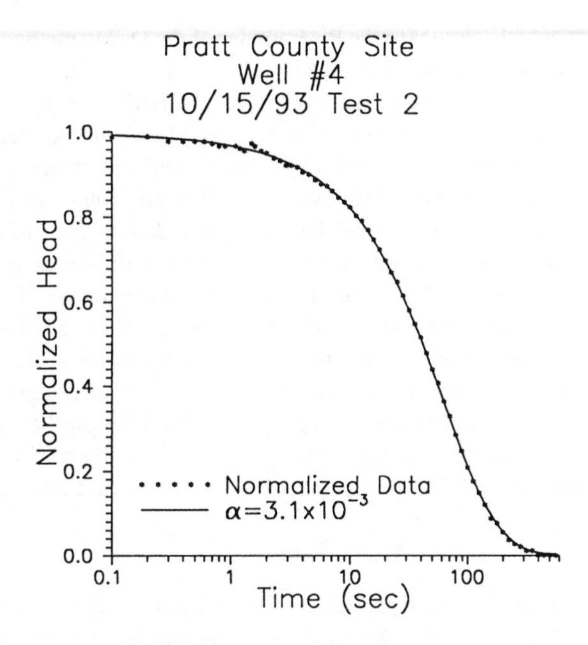

FIGURE 6.5 Normalized head ($H(t)/H_0$) vs. log time plot of the slug test in well 4 at the Pratt County site and the best-fit KGS Model type curve.

than that obtained with the KGS model. Substituting 3.1×10^{-3} for α_{cal} in Equation (6.9b) yields an estimate for specific storage of 5.3×10^{-4} m^{-1}, a value that is quite reasonable for alluvial deposits (Domenico and Schwartz, 1990). It is important to note, however, that response data for this test are quite insensitive to α. Butler et al. (1993) show that repeating the analysis with an α that is 161 times smaller than that found here yields a K_r estimate that is only 14% larger than the value obtained in this analysis.

Additional Methods

The vast majority of slug tests performed in wells screened below the water table are analyzed using one of the preceding three approaches. However, there is one other class of analysis methods that has been cited periodically in the groundwater literature. These approaches have been developed by agricultural engineers and soil scientists for the analysis of data from the piezometer test, the agricultural engineering equivalent of a slug test performed in a well screened below the water table. Piezometer-test methods are briefly described in this section.

A piezometer test is performed in a shallow cased hole with an open pit at the bottom. Thus, unlike tests in screened wells, water can flow into the well through the open bottom of the pit. The mathematical model upon which the techniques for the analysis of piezometer-test data are based is essentially the same as that underlying the Dagan method. Luthin and Kirkham (1949) provide an analytical solution to this mathematical model. Their solution results in an expression for the calculation of hydraulic conductivity that requires the estimation of a geometrical factor, termed a "shape" factor, which, similar to the P parameter of the Dagan method, is a function of the aspect ratio, the normalized distance below the water table, and the normalized distance above the lower boundary. Amoozegar and Warrick (1986) provide tabulated values of the shape factor calculated by Youngs (1968) using electrical analog simulations. The maximum aspect ratio for which tabulated values are provided is eight, an aspect ratio that is at the lower end of what would be found in most groundwater applications. Given that the well configuration (i.e., open bottom) is different from that normally employed in shallow groundwater investigations and that other methods are available for use over the same range of aspect ratios, approaches based on the piezometer method are not recommended for the analysis of slug tests performed in wells screened below the water table. Although several authors have stated that a series of piezometer tests can be designed so that the horizontal and vertical components of hydraulic conductivity can be estimated separately in anisotropic formations (e.g., Amoozegar and Warrick, 1986), existing analytical solutions (e.g., Dachler, 1935; Hyder et al., 1994) do not support those claims.

Wells Screened across the Water Table

The vast majority of slug tests performed for environmental applications in wells screened across the water table are analyzed with the Bouwer and Rice method. This method, however, was developed for the analysis of tests performed in wells in which the water table is above the top of the screen. Although not in common use, an analysis technique for slug tests in wells screened across the water table was proposed by Dagan (1978) for wells with aspect ratios greater than 50, ratios on the high end of those commonly used in shallow groundwater investigations. Agricultural engineers and soil scientists have developed techniques similar to that of Dagan for the analysis of data from auger-hole tests, the agricultural-engineering equivalent of a slug test performed in a well screened across the water table. Using the results of the work on auger-hole tests, the method of Dagan can be extended to wells of all common aspect ratios. The Bouwer and Rice method and the extension of the Dagan method are described in the following sections.

The Bouwer and Rice Method

The Bouwer and Rice method is based on the mathematical model defined by Equations (6.1a) to (6.1h). When applied to data from a test in a well screened across the water table, the steps employed in the analysis are exactly the same as those used for a well screened below the water table.

It is important to emphasize that when the Bouwer and Rice method is applied to data from a test in a well screened across the water table, the analyst is adopting a simplified representation of the flow system, i.e., both the position of the water table and the effective screen length, are not changing significantly during the course of the test. As described in Chapter 2, a series of slug tests can be designed to assess the appropriateness of this conceptualization. If the results of the series of tests indicate that this simplified representation of the flow system is a reasonable approximation, the Bouwer and Rice method, or any of the other techniques described in the previous sections of this chapter, can be employed to analyze the response data. If not, the Dagan method described in the following section should be used.

The empirical coefficients used in the Bouwer and Rice method were developed for slug tests in which the effective screen length does not change. However, as long as H_0 is small relative to the effective screen length at static conditions (<25% would be a conservative definition of "small"), the error introduced by changes in the effective screen length is quite small. Note that the original relationships developed by Bouwer and Rice (1976) for the estimation of the empirical coefficients included an extrapolation to the water table as a result of a limitation in their electrical analog model. Hyder and Butler (1995) have shown, however, that the error introduced by this extrapolation is of no practical significance.

It is not uncommon for response data from slug tests in wells screened across the water table to display a concave-upward curvature when plotted as the logarithm of head vs. time (e.g., Figure 6.6), especially in wells with artificial filter packs screened in formations of moderate to low hydraulic conductivity. Although this concave-upward curvature can be a product of a large α as discussed in Chapter 5, Bouwer (1989) speculated that response data displaying this curvature, which he termed the double straight line effect, are a result of a rapid draining of the filter pack (represented by interval A-B in Figure 6.6 according to his hypothesis) followed by a much slower response controlled by the hydraulic conductivity of the formation (represented by interval B-C in Figure 6.6). Bouwer carefully qualified his hypothesis by stating that such behavior should only be seen when the filter pack intersects the water table and is considerably more permeable than the formation. In this situation, he recommended that the Bouwer and Rice method be modified by using a slope estimate obtained from a straight line fit to the second linear segment (interval B-C of Figure 6.6) and an effective casing radius (r_c) given by the following formula:

$$r_c = \left[r_{nc}^2 + n\left(r_w^2 - r_{nc}^2\right)\right]^{0.5} \tag{6.10}$$

where: n = drainable porosity of the filter pack, [dimensionless];
r_{nc} = nominal radius of the well screen, [L];
r_w = outer radius of the filter pack, [L].

Error can be introduced into the r_c parameter, and thus the K_r estimate, through uncertainty about the porosity and the outer radius of the filter pack. A more appropriate manner to estimate r_c is to use a mass balance similar to that described in Chapter 3. Equation (3.1) can be rewritten for the case of a well screened across the water table as:

$$H_0^* \pi r_{nc}^2 = H_0^+ \pi\left(r_{nc}^2 + n\left(r_w^2 - r_{nc}^2\right)\right) = H_0^+ \pi r_c^2 \tag{6.11a}$$

where: H_0^* = the expected magnitude of the initial displacement in a well with nominal radius of r_{nc};

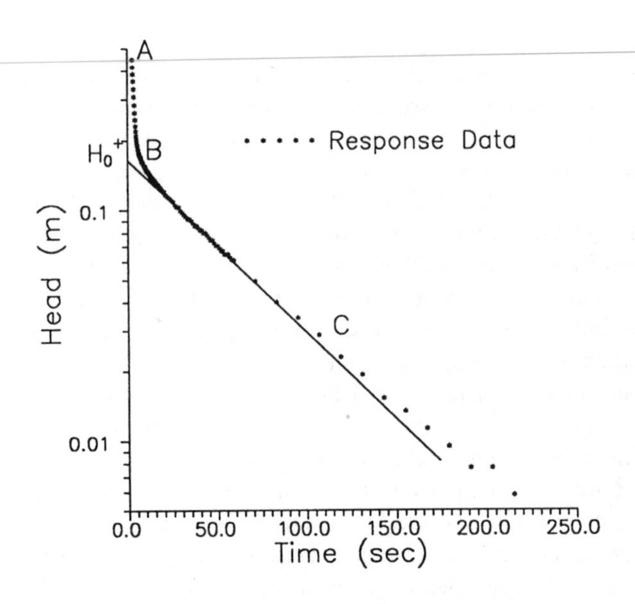

FIGURE 6.6 Logarithm of head vs. time plot illustrating the double straight line effect. A, B, C, and H_0^+ are defined in text. (Data courtesy of G. Zemansky.)

H_0^+ = the apparent magnitude of the initial displacement estimated from the y intercept of the fitted straight line as shown in Figure 6.6.

Equation (6.11a) can be rewritten in terms of r_c as:

$$r_c = r_{nc}\sqrt{\frac{H_0^*}{H_0^+}} \tag{6.11b}$$

The r_c value is then used in Equation (6.3) for the estimation of K_r in the same manner as would be done for a test in a well screened below the water table.

The recommendation of Bouwer to base analyses on the second linear interval of the response plot (the double straight line analysis) has been widely applied by field practitioners, often with little regard to whether the well screen or filter pack intersects the water table. However, Butler (1996) has shown that the double straight line analysis can also give very reasonable K_r estimates in cases where the concave-upward curvature is produced by a large α, if the appropriate range of normalized heads is employed in the analysis. Thus, the double straight line method appears to be a reasonable approach for the analysis of data with a pronounced concave-upward curvature, regardless of the mechanisms producing the curvature. In cases where the concave-upward curvature is produced by a large α, and not drainage of the filter pack, the r_c estimate should be based on the H_0 value determined from test data obtained immediately after test initiation, and not on the y-intercept (H_0^+) of the fitted line.

TABLE 6.5
Well Construction Information for Burcham Park Monitoring Well

Well Designation	r_{nc} (m)	b (m)	B (m)
Burcham Park Monitoring Well	0.051	1.51	>10.67

As with the methods developed for tests in wells screened below the water table, uncertainty about anisotropy and the impact of noninstantaneous slug introduction are also issues of practical concern. The presentation on the Bouwer and Rice method for wells screened below the water table provides further discussion of the practical ramifications of these issues.

An example can be used to illustrate the application of the Bouwer and Rice method for the analysis of slug tests performed in wells screened across the water table. In October of 1996, a series of slug tests were performed in a shallow monitoring well located within Burcham Park in Douglas County, Kansas (Stanford et al., 1996). The well was screened in unconsolidated alluvial deposits consisting primarily of interbedded fine sands and silts. Table 6.5 summarizes the well-construction information, while Table 6.6 lists the test data employed in the analysis. Repeat tests using H_0 less than 20% of the effective screen length at static conditions indicated that changes in the effective screen length during a test had little, if any, impact on the response data. Figure 6.7A is a conventional log normalized head vs. time plot, illustrating the pronounced concave-upward curvature seen in tests at this site. In this case, the concave-upward curvature was interpreted to be a product of a large α, and not drainage of the filter pack. An r_c estimate of 0.051 m, the nominal screen radius, was calculated using Equation (3.1) and the H_0 value determined from data obtained immediately after test initiation. Figure 6.7B displays a straight line fit to the normalized head range recommended by Butler (1996). The T_0 value calculated from this straight line is 373.7 sec. Given the well construction information of Table 6.5, a value of 2.101 was calculated for the $\ln(R_e/r_w^*)$ term. Substitution of this value and the r_c and T_0 estimates into Equation (6.3) results in a K_r estimate of 0.41 m/d. In this case, the details of the fitting procedure have a considerable impact on the results because of the curvature displayed in Figure 6.7A. For example, utilization of an automated fitting procedure that emphasizes the early portions of the data record yields a hydraulic conductivity estimate of 0.78 m/d, a value that is nearly twice as large as that obtained using the normalized head range (0.20 to 0.30) recommended by Butler (1996).

The Dagan Method

The Dagan method for the analysis of slug tests performed in wells screened across the water table (the response test in the terminology of Dagan [1978]) is based on a mathematical model defined for the general anisotropic case as:

TABLE 6.6
Response Data from 10/01/96 Slug Test #2 in Burcham Park Monitoring Well

Time (s)	Head (m)	Normalized Head
0.1	0.261	1.019
0.2	0.247	0.964
0.3	0.248	0.969
0.4	0.258	1.008
0.5	0.245	0.957
0.6	0.252	0.985
0.7	0.248	0.971
0.8	0.249	0.972
0.9	0.248	0.970
1.0	0.247	0.966
1.2	0.247	0.967
1.4	0.247	0.966
1.6	0.247	0.964
1.9	0.245	0.958
2.2	0.245	0.956
2.6	0.243	0.948
3.0	0.242	0.945
3.5	0.241	0.940
4.0	0.240	0.937
4.7	0.239	0.934
5.5	0.238	0.928
6.4	0.237	0.925
7.4	0.234	0.914
8.6	0.233	0.910
10.0	0.231	0.902
11.7	0.227	0.888
13.6	0.224	0.876
15.9	0.221	0.865
18.5	0.218	0.853
21.6	0.215	0.839
25.2	0.209	0.817
29.3	0.206	0.805
34.2	0.200	0.781
39.9	0.196	0.766
46.5	0.188	0.736
54.2	0.183	0.714
63.1	0.175	0.682
73.6	0.165	0.643
85.8	0.156	0.608
100.0	0.148	0.578
116.6	0.136	0.531
136.0	0.125	0.490

TABLE 6.6
Response Data from 10/01/96 Slug Test #2 in Burcham Park Monitoring Well

Time (s)	Head (m)	Normalized Head
158.5	0.115	0.448
184.8	0.103	0.403
215.5	0.092	0.358
251.2	0.084	0.328
292.9	0.072	0.279
341.5	0.065	0.253
398.2	0.056	0.218
464.2	0.047	0.182
541.2	0.039	0.151
631.0	0.033	0.127
735.7	0.029	0.113
857.7	0.023	0.091
1000.0	0.016	0.061
1166.0	0.015	0.058

$$\frac{\partial^2 h}{\partial r^2} + \frac{1}{r}\frac{\partial h}{\partial r} + \frac{K_z}{K_r}\frac{\partial^2 h}{\partial z^2} = 0 \tag{6.12a}$$

$$H(0) = H_0 \tag{6.12b}$$

$$h(r, 0, t) = 0,\ r_w < r < \infty,\ t > 0 \tag{6.12c}$$

$$\frac{\partial h(r, B, t)}{\partial z} = 0,\ r_w < r < \infty,\ t > 0 \tag{6.12d}$$

$$\frac{\partial h(\infty, z, t)}{\partial r} = 0,\ 0 \le z \le B,\ t > 0 \tag{6.12e}$$

$$h(r_w, z, t) = z,\ 0 \le z < H(t),\ t > 0 \tag{6.12f}$$

$$h(r_w, z, t) = H(t),\ H(t) \le z \le b,\ t > 0 \tag{6.12g}$$

$$2\pi r_w K_r \int_0^b \frac{\partial h(r_w, z, t)}{\partial r} dz = \pi r_c^2 \frac{dH(t)}{dt},\ t > 0 \tag{6.12h}$$

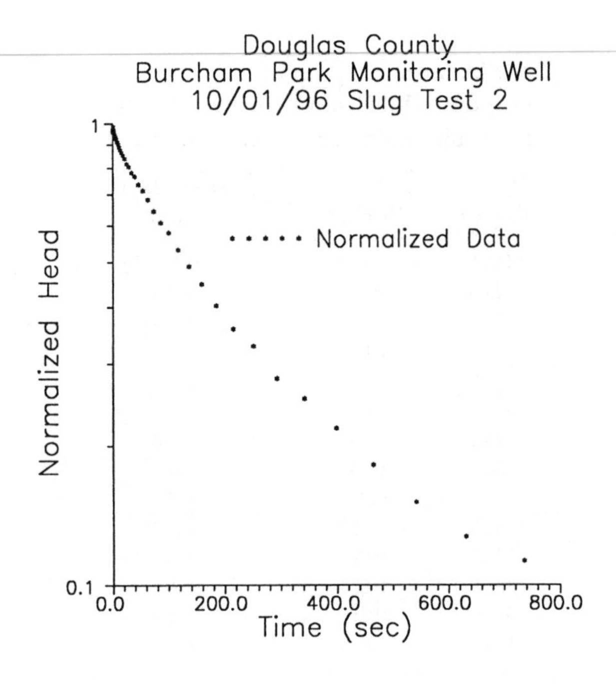

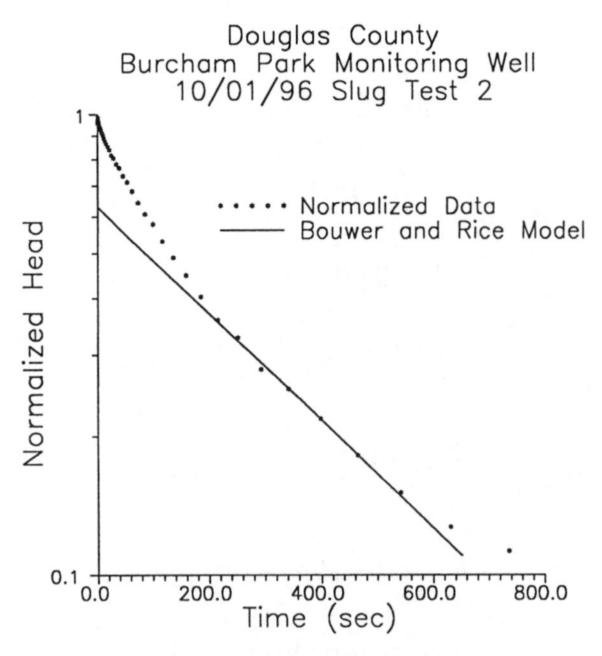

FIGURE 6.7 (A) Logarithm of normalized head ($H(t)/H_0$) vs. time plot of a slug test performed in the Burcham Park monitoring well in Douglas County, Kansas; (B) Logarithm of normalized head ($H(t)/H_0$) vs. time plot of the Burcham Park test and the Bouwer and Rice model fit for the normalized head range recommended by Butler (1996).

$$\frac{\partial h(r_w, z, t)}{\partial r} = 0, b < z \leq B, t > 0 \tag{6.12i}$$

Note that z equals zero at the water table and increases in a downward direction in this model.

Boast and Kirkham (1971) propose an analytical solution for a mathematical model similar to that defined by Equations (6.12a) to (6.12i) and evaluate the solution for aspect ratios between 1 and 100. Dagan (1978) employs a more efficient approximate semianalytical approach to evaluate that solution for aspect ratios greater than 50. Dagan shows how the solution for all aspect ratios can be written as:

$$\ln\left(\frac{H(t)}{H_0(2b - H(t))/(2b - H_0)}\right) = -\frac{2K_r bt}{(1/P)r_c^2} \tag{6.13}$$

where: P = dimensionless flow parameter.

When H_0 is small with respect to 2b, Equation (6.13) reduces to the same form as used in the methods of Dagan (1978) and Bouwer and Rice (1976) for tests in wells screened below the water table.

The form of Equation (6.13) indicates that a plot of the logarithm of the normalized head term on the left-hand side vs. time will be linear. Thus, analogous to the techniques discussed earlier, the Dagan method for wells screened across the water table involves calculating the slope of a straight line fit to a plot of the response data and then using that value to estimate the radial component of hydraulic conductivity. The method essentially consists of the following six steps:

1. The response data are plotted as the logarithm of the left-hand side of Equation (6.13) vs. the time since test initiation as shown in Figure 6.8A;
2. A straight line is fit to the data plot either via visual inspection or an automated regression routine;
3. The slope of the fitted line is calculated. If a criterion analogous to the time lag (T_0) is used in this calculation, the slope, when written in terms of the natural logarithm, becomes ($-1/T_0^*$), where T_0^* is the time at which the plotted logarithmic term is equal to 0.368;
4. An estimate of the ψ parameter for the particular well-formation configuration is obtained. In most cases, a value of one is used for the anisotropy ratio;
5. Given the ψ value and the normalized length of the well screen (b/B), a value for P, the dimensionless flow parameter, is calculated;
6. The radial component of hydraulic conductivity is estimated using the following equation:

$$K_r = \frac{r_c^2(1/P)}{2bT_0^*} \tag{6.14}$$

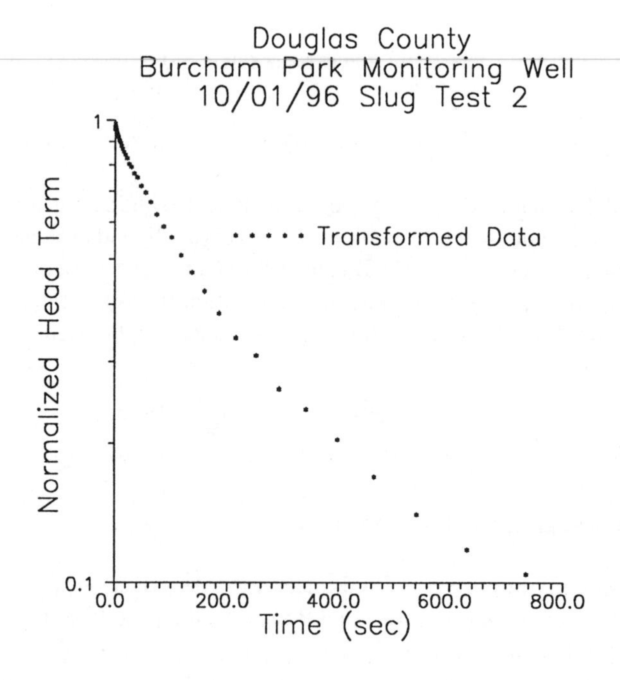

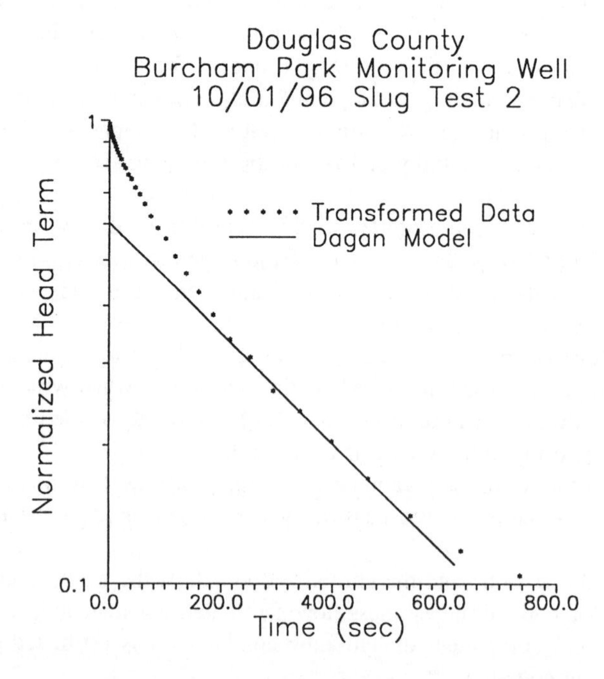

FIGURE 6.8 (A) Logarithm of normalized head term of Equation (6.13) vs. time plot of a slug test performed in the Burcham Park monitoring well in Douglas County, Kansas; (B) Logarithm of normalized head term vs. time plot of the Burcham Park test and the Dagan model fit for the normalized head range recommended by Butler (1996).

Note that Equation (6.14) has the same general form as the expressions used in all other analysis methods based on the quasi-steady-state representation of slug-induced flow.

There are several issues of practical importance with respect to the Dagan method. The most significant of these is the estimation of the P parameter in Equation (6.14). Dagan (1978) provides P values for selected aspect ratios greater than 100 ($\psi \leq 0.01$). Dagan cites the approximate formulae of Ernst (Bower and Jackson, 1974) as the source of P values over a wider range of aspect ratios. The Ernst formula for the case of a very small normalized screen length (b/B approaches zero) can be rewritten in the notation of this book as:

$$P = 0.216(1 + 20\psi) \tag{6.15}$$

The tabulated P values presented by Dagan, however, indicate that Equation (6.15) should not be applied for aspect ratios much greater than 100 (ψ values much less than 0.01). Since the Ernst formulae have only been developed for normalized screen lengths approaching one and zero, the difference in P values between these two end-member cases can be quite large for wells of moderate to small aspect ratios (moderate to large ψ values). For cases between these two end members, the solution of Boast and Kirkham (1971) should be used to compute P values. Boast and Kirkham present tabulated values of shape factors that can be converted to P values using the following equation:

$$P = \frac{864(r_c/b)^2}{2\frac{H(t)}{b}\left(1 - \frac{0.5H(t)}{b}\right)C_{BK}} \tag{6.16}$$

where: C_{BK} = shape factor of Boast and Kirkham for a particular H(t), b, and r_c, [dimensionless].

As a result of their proposed analysis approach, the C_{BK} parameter of Boast and Kirkham is a function of H(t), a dependence that makes the analysis procedure more complicated. However, the P parameter employed in the Dagan method is not dependent on H(t). Table 6.7 displays selected P values that were calculated from the table of Boast and Kirkham (1971) using Equation (6.16). Table 6.8 compares the P values calculated using the Boast and Kirkham solution for the case of a normalized screen length approaching zero with values obtained via approximate approaches. The column labelled Ernst in Table 6.8 provides estimates obtained using the approximate formula of Ernst (Equation [6.15]). A comparison with the column of Boast and Kirkham values indicates that the Ernst equation will provide P estimates within 12% of the Boast and Kirkham values for aspect ratios between 5 and 100. Thus, the Ernst equation should be employed for the estimation of P whenever the normalized screen length is less than 0.5. The column labelled Bouwer and Rice in Table 6.8 provides values for the $(1/(\ln(R_e/r_w^*))$ parameter, which is

TABLE 6.7
Tabulated Values of the Dimensionless Flow Parameter, P, Used in Dagan Method for Wells Screened Across the Water Table

	b/B					
ψ	1.0	0.91	0.83	0.67	0.5	≈0.0
0.20	0.666	0.748	0.807	0.893	0.937	0.965
0.10	0.477	0.537	0.572	0.613	0.635	0.645
0.050	0.365	0.408	0.427	0.449	0.460	0.466
0.020	0.276	0.303	0.311	0.323	0.329	0.332
0.010	0.234	0.254	0.254	0.262	0.270	0.270

Note: Values generated with the analytical solution of Boast and Kirkham (1971).

TABLE 6.8
Comparison of P Values Computed with Boast and Kirkham (1971) Solution and Values Obtained with Approximate Approaches

ψ	Boast and Kirkham	Ernst	Bouwer and Rice	Cole and Zlotnik
0.20	0.965	1.080	1.264	1.045
0.10	0.645	0.648	0.827	0.687
0.050	0.466	0.432	0.582	0.491
0.020	0.332	0.302	0.418	0.342
0.010	0.270	0.259	0.346	0.276

Note: Cole and Zlotnik values are from (d+b)/b = 1.05 Column of Table 6.3.

analogous to P, used in the Bouwer and Rice method for the case of a normalized screen length approaching zero. In this range of aspect ratios, the empirical parameter employed in the Bouwer and Rice method is approximately 25-30% larger than the Boast and Kirkham value. This 25-30% overestimation of P will lead to K_r estimates obtained with the Bouwer and Rice method that underpredict the actual value by 20-25%, a finding consistent with those of Dagan (1978) and Hyder and Butler (1995) for a well screened below the water table. The final column in Table 6.8, labelled Cole and Zlotnik, provides P values computed using the semianalytical solution of Cole and Zlotnik (Cole and Zlotnik, 1994; Cole, pers. commun., 1996) for the case of a well screened just below the water table (d approaching zero). As would be expected, the solution of Cole and Zlotnik is in close agreement with that of Boast and Kirkham.

A second issue of practical concern is the difference between the well configuration assumed in the mathematical model defined by Equations (6.12a) to (6.12i) and that assumed in the model of Boast and Kirkham (1971). As in the case of the mathematical models developed for the analysis of piezometer tests, the model of Boast and Kirkham assumes that water can flow upward through the bottom of the well. However, the close agreement between the various columns of Table 6.8 indicates that the magnitude of this vertical flow is quite small for the range of aspect ratios used in slug tests for environmental applications.

All of the mathematical models discussed in this book until now have been linear in form. The linear nature of the underlying mathematical models has been reflected in the solutions for normalized head, which indicate that the normalized response data are independent of the magnitude of H_0. When these linear models are appropriate representations of the underlying physics, plots of normalized response data from a series of tests initiated with different H_0 values will coincide. However, as a result of a nonlinear boundary condition along the well screen (Equations [6.12f] and [6.12g]), the solution given in Equation (6.13) does not retain this independence from H_0. Thus, when H_0 is a significant fraction of the effective screen length at static conditions (above at least 0.25 [Stanford et al., 1996]), plots of the normalized response data will no longer coincide. Instead, the larger the H_0, the greater the test duration. This head dependence will also produce a pronounced curvature when the logarithm of the normalized head is plotted vs. time (e.g., Figure 2.8). Although plots of the response data normalized in the conventional fashion will no longer coincide when the mathematical model defined by Equations (6.12a) to (6.12i) is a reasonable representation of the governing physics, the responses will coincide when the data are plotted as the logarithmic term on the left-hand side of Equation (6.13) vs. time. Thus, if a reproducible dependence on H_0 is observed in response data from a well screened across the water table, a plot in the format given in Equation (6.13) will indicate if the dependence is a product of the nonlinearity introduced by changes in the effective screen length during the course of a test. If the nonlinearity is deemed of significance, then the analysis method outlined in this section should be employed. If the nonlinearity is small to nonexistent, the Bouwer and Rice method and other approaches proposed for the analysis of tests in wells screened below the water table should be considered as equally viable alternatives to the method described in this section.

The Burcham Park slug test described in the previous section can be used to demonstrate the Dagan method for the analysis of slug tests performed in wells screened across the water table. Figure 6.8A is a plot of the response data transformed according to the left-hand side of Equation (6.13). The pronounced concave-upward curvature exhibited by the response data is interpreted to be a product of a large α. Figure 6.9 displays both a plot of the conventional normalized head and a plot of the transformed data. The similarity of these plots and their pronounced concave-upward curvature are strong evidence that the effect of the nonlinear mechanisms (i.e., changes in the effective screen length) is quite small for this test. Figure 6.8B displays a straight line fit to the normalized head range recommended by Butler (1996). The T_0^* estimate obtained from this straight line is 364.0 sec. Since the lower impermeable boundary is at least 11 m below the water table, Table 6.7 indicates that a normalized screen length of 0 can be used for this case. Given an aspect ratio of 30, the ψ value is 0.033. Using the Ernst formula given in

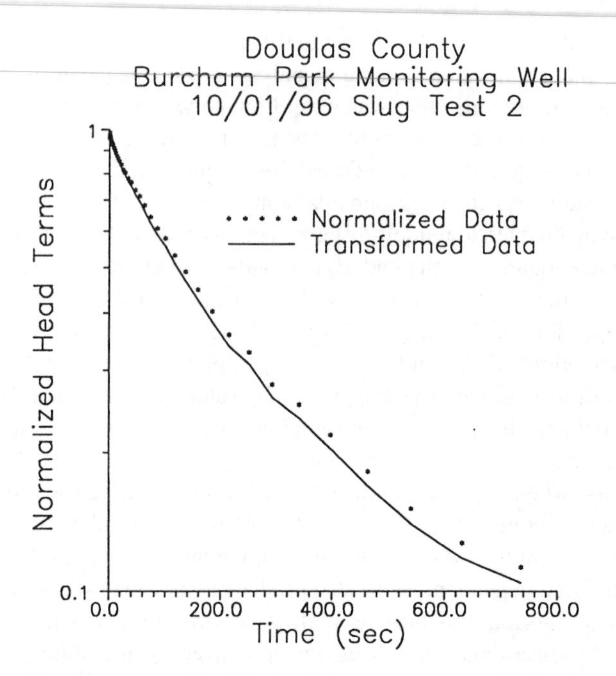

FIGURE 6.9 Logarithm of normalized head ($H(t)/H_0$) and normalized head term of Equation (6.13) vs. time plots for the Burcham Park test (response data transformed according to Equation (6.13) are plotted as solid line).

Equation (6.15), a value of 0.359 is calculated for P. Substitution of this value and the T_0^* estimate into Equation (6.14) yields a hydraulic conductivity estimate of 0.56 m/d, a value that is approximately 36% above the Bouwer and Rice estimate obtained in the previous section. Note that the results of the comparison of the two K_r estimates are in agreement with the theoretical findings discussed earlier with respect to Table 6.8.

7 The Analysis of Slug Tests — Low Conductivity Formations

CHAPTER OVERVIEW

The analysis methods discussed in Chapters 5 and 6 were developed without any consideration of the hydraulic conductivity of the formation. As long as the underlying mathematical models are appropriate representations of the governing physics, these techniques can be used in formations of any hydraulic conductivity. However, pragmatic considerations, specifically the length of time necessary to complete a test, may make conventional approaches for the performance and analysis of slug tests of rather limited use in low-conductivity formations. In an effort to make the slug test a more viable method for this setting, specialized test techniques have been developed. The most common of these techniques are described in this chapter. Additional considerations for tests in formations of extremely low hydraulic conductivity are also briefly summarized.

SLUG TESTS IN LOW-CONDUCTIVITY FORMATIONS

The duration of a slug test is a function of both formation and well-construction parameters. It is possible to decrease the duration of a test in low-permeability materials by modifying the well-construction parameters. Although the effective radius of the well screen does have a small effect on test duration, the well parameters that have the largest impact are the effective screen length and casing radius. Whether a test is performed in a fully or partially penetrating well, the major factors affecting test duration are encapsulated in the β parameter, which for a partially penetrating well is defined as

$$\beta = \frac{\left(K_r bt\right)}{r_c^2} \tag{7.1}$$

The form of Equation (7.1) indicates that test duration is inversely proportional to the effective screen length (b) and directly proportional to the square of the effective casing radius (r_c). Since changes in the casing radius will have the largest impact on test duration, as a result of the squared term in Equation (7.1), and are relatively simple to effect in practice, most efforts to decrease the duration of a slug test have focused on decreasing the effective radius of the well casing.

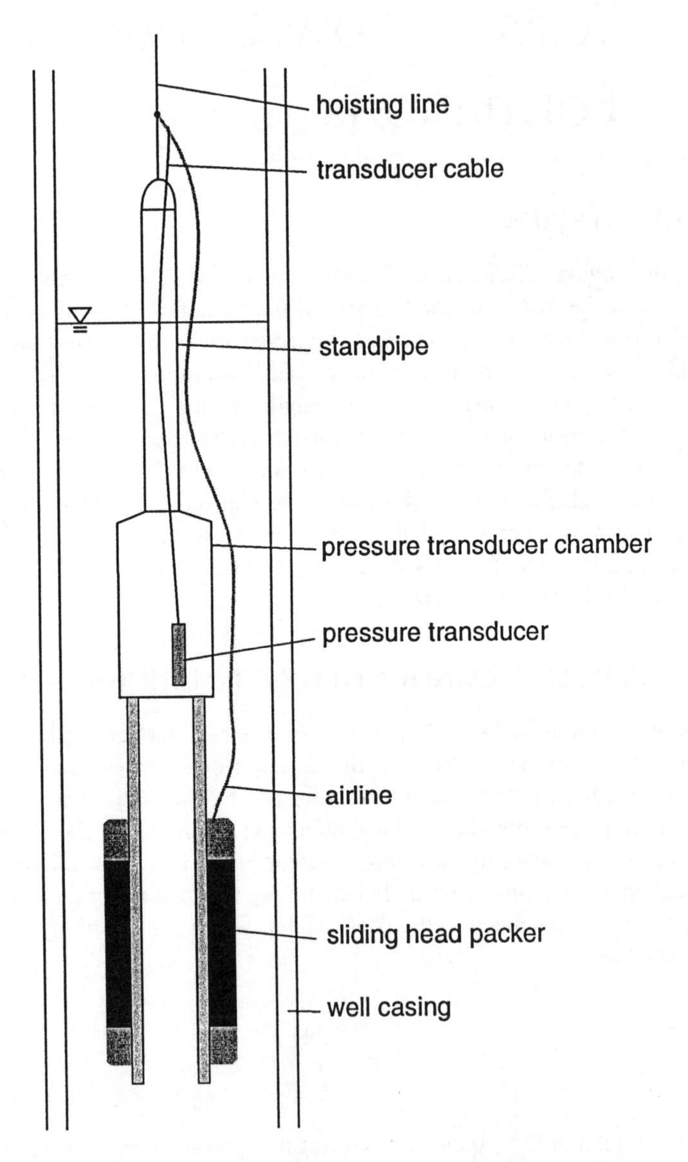

FIGURE 7.1 Schematic of a packer and standpipe apparatus used to decrease the effective radius of the casing (water level changes during a test occur only in smaller diameter standpipe; apparatus shown with uninflated packer; figure not to scale).

The simplest way to decrease the effective casing radius is to use a packer and standpipe arrangement such as that shown in Figure 7.1. Figure 7.2 shows how test duration was decreased at a well in Reno County, Kansas by using the packer and

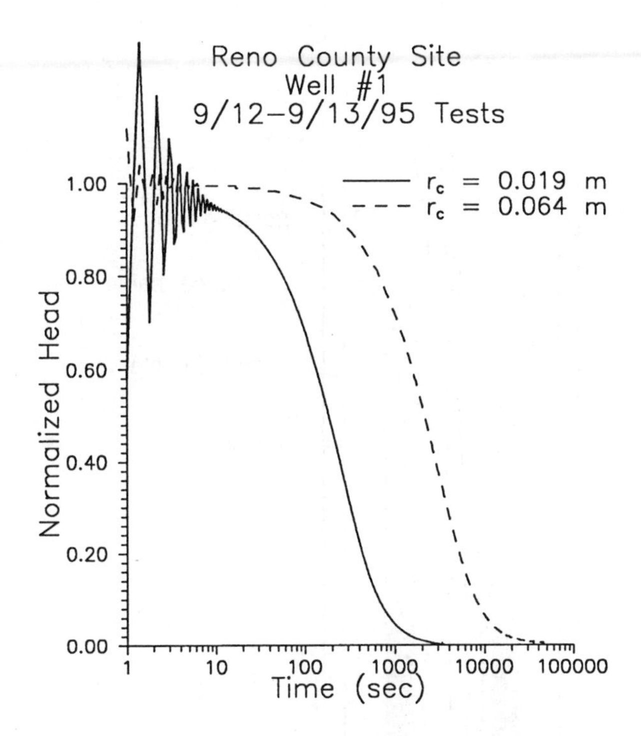

FIGURE 7.2 Normalized head ($H(t)/H_0$, where H(t) is deviation from static and H_0 is magnitude of the initial displacement) vs. log time plot of a pair of slug tests performed in well 1 at a Reno County, Kansas monitoring site using different effective casing radii (nominal radius of well casing is 0.064 m; early time fluctuations related to test initiation).

standpipe method. At this well, test duration, as measured by T_0, was decreased by a factor of 12.4 (from 3274 s to 265 s) when a 0.019 m radius standpipe was used in a well with a nominal casing radius of 0.064 m. Given the casing normally used in shallow groundwater applications (r_c = 0.025 to 0.076 m), this approach can be readily employed to obtain decreases in duration on the order of a factor of 4 to a factor of 20 or more.

If the interval is of very low hydraulic conductivity, the standpipe radius required to produce a test of a reasonable duration may be so small as to be impractical. Bredehoeft and Papadopulos (1980) describe a hypothetical slug test in a formation with a hydraulic conductivity of 8.64×10^{-8} m/d. In their example, a standpipe radius of 0.0008 m would be required to complete a test within 1 day. It is probably not practical, however, to consider using standpipes much smaller than 0.005 m in radius. For conditions that would require smaller standpipes, Bredehoeft and Papadopulos (1980) propose a modified slug-test procedure that will be designated here as the "shut-in" slug test. Although primarily utilized by petroleum engineers, a related approach, known as the drillstem test, has also been applied in hydrogeologic investigations. These two techniques are described in the following sections, after which additional considerations for tests in formations of extremely low hydraulic conductivity are discussed.

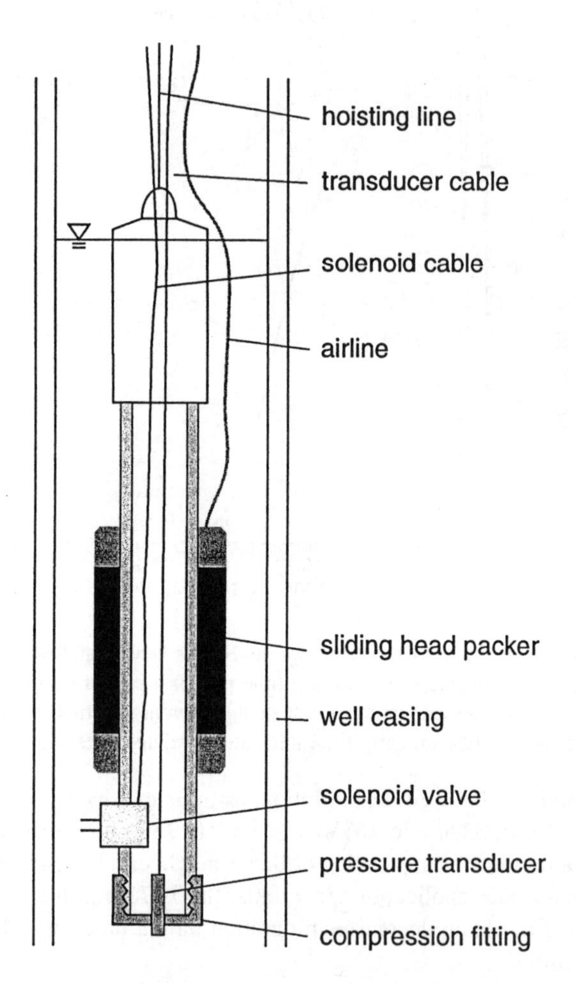

FIGURE 7.3 Schematic of apparatus used for shut-in tests (transducer measures pressure below packer; valve activated from the surface; apparatus shown with uninflated packer; figure not to scale).

The Shut-In Slug Test

In a conventional slug test, the head will change in the screened interval as the water level changes in the open casing in response to flow from/to the well:

$$\pi r_c^2 \frac{dH}{dt} = 2\pi r_w K_r b \frac{\partial h}{\partial r} \tag{7.2a}$$

The shut-in slug test is based on isolating the screened interval of the well from the portion of the casing open to the atmosphere. This is commonly done, as depicted in Figure 7.3, by using an inflatable packer in conjunction with a valve that can be opened and shut from the surface. When the screened interval is isolated from the open casing, head changes are no longer a function of water level changes in the casing. Instead, the head changes are a function of the compressibility of water and the test equipment (packer and associated tubing). In this case, Equation (7.2a) can be rewritten as:

$$V_w C_{ef} \rho_w g \frac{dH}{dt} = 2\pi r_w K_r b \frac{\partial h}{\partial r} \tag{7.2b}$$

where: V_w = volume of water in the well below the packer, $[L^3]$;
C_{ef} = effective compressibility of the water below the packer, includes the compressibility of both water and test equipment, $[LT^2/M]$;
ρ_w = density of water, $[M/L^3]$;
g = gravitational acceleration, $[L/T^2]$.

Neuzil (1982) emphasizes the need to include the compressibility of the test equipment in the effective compressibility term, and presents an example in which the effective compressibility is approximately six times larger than the compressibility of water. Note that Kell (1975) reports a value of 4.7×10^{-10} Pa^{-1} for the compressibility of water at 13°C, a temperature that is a reasonable average value for shallow groundwater.

In order to compare conventional and shut-in slug tests performed in the same well, it is useful to compute an effective casing radius for the shut-in test. This can be done by rewriting Equation (7.2b) in the form of Equation (7.2a), and defining the effective casing radius of a shut-in test (r_c') as:

$$r_c' = \left(\left(V_w C_{ef} \rho_w g \right) / \pi \right)^{0.5} \tag{7.3}$$

Although they ignore equipment compressibility, Bredehoeft and Papadopulos (1980) use an equation similar to (7.3) to compute an effective casing radius of 0.0008 m for the hypothetical slug test they describe.

A series of shut-in slug tests were performed in the Permian bedrock underlying the Great Bend Prairie aquifer of southcentral Kansas in November of 1994 using the apparatus depicted in Figure 7.3. One of the wells tested in this series was the Reno County well discussed earlier in this chapter (Figure 7.2). Assuming negligible equipment compressibility, an effective casing radius of 0.0009 m ($V_w = 0.61$ m^3) was calculated for the Reno County well. Figure 7.4 displays the record of the three shut-in tests performed at this well. Note that the response data for the first test (test A) lag those of the following tests (T_0 50% larger for test A) as a result of heads not being at static conditions prior to initiation of the first test. A similar lag was seen at all of the wells tested in November of 1994.

Figure 7.5 is a plot of conventional and shut-in slug tests performed in the Reno County monitoring well. A comparison of T_0 values indicates that test duration was decreased by a factor of approximately 20 between the test performed with the

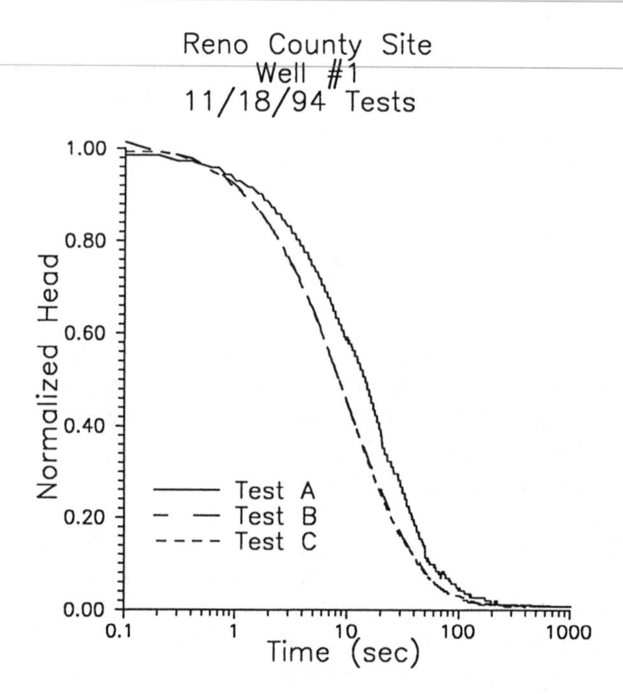

FIGURE 7.4 Normalized head ($H(t)/H_0$) vs. log time plot of three slug tests performed in well 1 at the Reno County, Kansas monitoring site using the shut-in test procedure with the apparatus shown in Figure 7.3 (test A was first test in sequence).

0.019 m radius standpipe ($T_0 = 265$ s) and that performed with the shut-in procedure ($T_0 = 13.3$ s). This decrease, however, was much less than what was expected from a calculated r_c' of 0.0009 m. The difference between the expected and actual durations for the shut-in tests is most likely a result of equipment compressibility (Neuzil, 1982). A reasonable estimate of the effective compressibility can be obtained by comparing results from tests performed in the same well using different effective casing radii. At the Reno County well, slug tests were performed using three different effective casing radii. A hydraulic-conductivity estimate of 0.092 m/d can be obtained for the test performed with the 0.019 m standpipe using the Hvorslev method. Assuming that this conductivity estimate is also appropriate for the shut-in test, a value of 0.0055 m can be calculated for r_c'. A comparison with the original r_c' estimate indicates that the effective compressibility of the test interval was approximately 35 times larger than the compressibility of water. This very large difference is primarily a result of inattention to compressibility issues when constructing the test equipment. Note that this simple approach for the estimation of the effective compressibility should only be employed in wells, such as the Reno County monitoring well, that do not display strong dynamic skin effects.

The estimate of equipment compressibility obtained at the Reno County well was used for all the shut-in tests performed in the November 1994 test program. The hydraulic-conductivity estimates obtained from these tests were in reasonable agreement with those determined using other approaches. Decreases in duration of

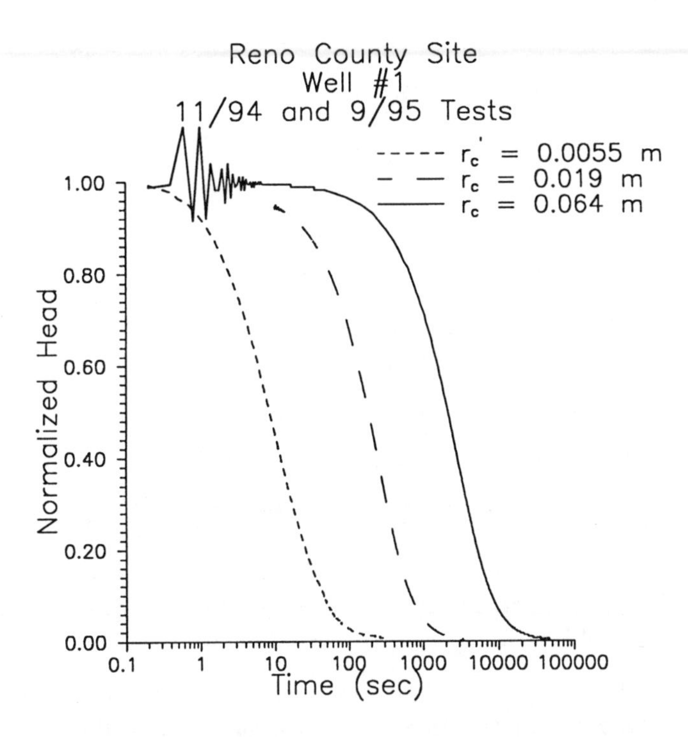

FIGURE 7.5 Normalized head ($H(t)/H_0$) vs. log time plot of three slug tests performed in well 1 at the Reno County, Kansas monitoring site using three different effective casing radii (nominal radius of well casing is 0.064 m; r_c of 0.019 m obtained with the apparatus of Figure 7.1; r_c' of 0.0055 m obtained with the apparatus of Figure 7.3).

a factor of 200 or more, with respect to slug tests performed in the nominal well casing, were obtained at all wells.

In addition to the consideration of equipment compressibility, there are several other issues of practical importance with respect to the shut-in slug test. The most important of these is the amount of flow into/out of a well that is necessary to produce a unit change in head. As the effective casing radius decreases, the amount of flow required to produce a unit change in head decreases as a function of r_c^2. Thus, in a shut-in test, only a very small volume of water moves into/out of the well in response to the slug-induced disturbance. As the amount of flow per unit change in head decreases, the volume of the formation affected by the test decreases as well. The result is that a shut-in test impacts a very small volume of the formation. In some cases, the affected volume may be so small that it does not extend beyond the zone of disturbance created during drilling and development (e.g., Moench and Hsieh, 1985a). Thus, considerable care must be used to interpret the results obtained from a slug test performed with the shut-in procedure.

A second issue of practical importance is that of test initiation prior to obtaining equilibrium conditions. In low-conductivity formations, the effect of drilling and development on formation heads can last for a considerable period for time. In addition, as was shown by the series of tests depicted in Figure 7.4, head disturbances

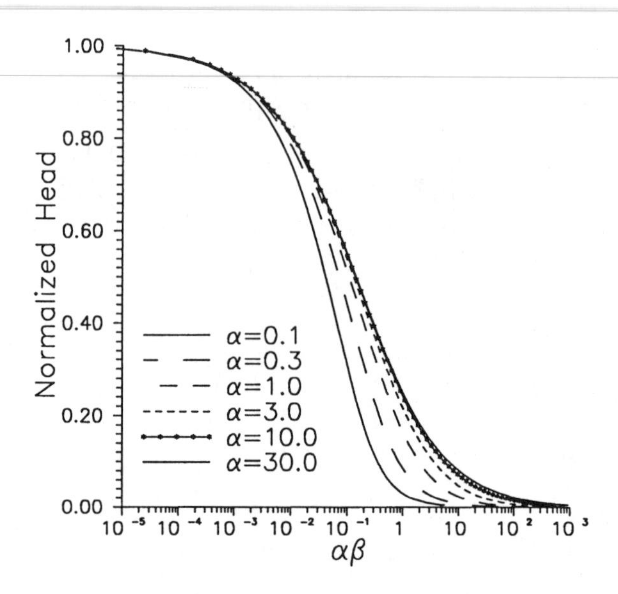

FIGURE 7.6 Normalized head ($H(t)/H_0$) vs. logarithm of $\alpha\beta$ type curves generated with the Cooper et al. model.

caused by emplacing equipment in the well and related test preparation can take considerable time to dissipate. Thus, as recommended by Neuzil (1982), the well should be shut in and heads allowed to reach near-equilibrium conditions prior to test initiation.

A third issue of practical importance is that of the nonuniqueness of the type-curve match obtained with the Cooper et al. or KGS models. When α is greater than about 0.1, considerable error may be introduced into the K_r estimate as a result of similarities in the shape of the type curves. Bredehoeft and Papadopulos (1980) suggest a modified type curve approach that involves fitting test data to type curves of normalized head vs. the logarithm of the product of α and β (Figure 7.6). An estimate of α can be obtained from the type curve match and then used to calculate β, and, thus, K_r from the match point on the x axis. As shown in Figure 7.6, α and β cannot be estimated separately for α values larger than five to ten. Fortunately, however, such values are rarely met in groundwater applications.

Additional practical issues of importance for shut-in tests include the impact of a leaky annular seal, entrapped air in the filter pack, and leaks at joints in the well casing. Palmer and Paul (1987) provide further discussion of these issues.

Note that the shut-in test has also been referred to as the "pulse test" (e.g., Connell, 1994) and the "pressurized slug test" (e.g., Guyonnet et al., 1993) in the groundwater literature. However, the pulse-test designation can lead to confusion as petroleum engineers use this term to describe a variable-rate procedure for conducting multiwell pumping tests (e.g., Johnson et al., 1966; Streltsova, 1988), while the pressurized slug-test designation can lead to confusion with tests initiated by pressurizing the air in the well casing (see Chapter 3).

THE DRILLSTEM TEST

The drillstem test is also used by hydrogeologists to estimate the hydraulic conductivity of low-permeability formations, especially in the case of formations that are at a considerable depth below the land surface (e.g, Beauheim, 1987; Karasaki, 1991). Although not frequently utilized in shallow groundwater investigations, the drillstem test is very commonly used for the evaluation of petroleum reservoirs. This technique essentially is a slug test that begins as a conventional open-hole test and then is shut-in at a relatively large normalized head (usually greater than 0.5). The technique is performed in the petroleum industry using an apparatus similar in concept to the packer and standpipe arrangement depicted in Figure 7.1, in which the packer has been modified as in Figure 7.3. The open-hole portion of the test involves water level changes in the standpipe, while the shut-in portion of the test involves head changes below the packer (the normalized head at which the well is shut-in is termed the shut-in head (h_s)). The technique is usually performed in a rising-head mode, i.e., water is removed from the standpipe for test initiation. In the petroleum industry, the test is initiated using a H_0 on the order of hundreds to thousands of meters of water.

The drillstem test (commonly abbreviated as dst) was initially developed by petroleum engineers in the 1920s to obtain samples of the reservoir fluid from the flow that came into the standpipe during the open-hole portion of the test. With the development of reliable pressure sensors, the purpose of the test was enlarged to include estimation of the permeability and the equilibrium (static) head of the formation. In the petroleum industry, drillstem tests are virtually always performed in pairs. The first test is designed to remove drilling mud from the formation and allow formation heads to return to equilibrium conditions, while the second test is the focus of the analysis. A considerable body of terminology has been developed by petroleum engineers to describe different phases of a drillstem test. In particular, the open-hole portion of the test is designated as the flow period, while the shut-in portion of the test is designated as the buildup or shut-in phase. In the 1950s, petroleum engineers developed a method for the estimation of permeability from dst data that is based on the work of Horner (1951). However, as Bredehoeft (1965) points out, the work of Horner and the analysis method based on it are simply restatements of the Theis (1935) method for the analysis of recovery data following a pumping test (Kruseman and de Ridder, 1990).

The Theis recovery method was originally applied to dst data because the flow into the well during the open-hole phase (flow period) of the test is much greater than that during the shut-in phase, i.e., a drillstem test can be represented as a period of pumpage (flow period) followed by a period of recovery (shut-in phase). Experimentation with data from actual drillstem tests showed that the data did follow the linear relationship that would be predicted from the Theis recovery analysis; so, this technique was widely adopted by petroleum engineers. Although the flow rate is virtually always decreasing during drillstem tests performed for hydrogeologic applications, most analysts assume a constant rate and use the average rate during the flow period for this quantity. However, as Ramey et al. (1975) point out, the assumption of a constant rate during the flow period is quite good when the velocity of flow into the well is greater than the sonic (acoustic) velocity. This situation is most likely to arise when H_0 is extremely large (as is the norm in the petroleum industry) and the formation is of moderate or higher permeability. A constant rate during the flow period is not common in drillstem tests performed for hydrogeologic investigations, where the primary application is in low-permeability formations.

The early work on methods for the analysis of dst data was all done in the absence of a rigorous analytical solution to the mathematical model describing the slug-induced flow. The first analytical solution for a drillstem test, which uses a step function to represent the change in the effective casing radius, was described in the petroleum literature by Correa and Ramey (e.g., Correa and Ramey, 1987). Karasaki (1990) later introduced a solution in the groundwater literature that was derived using Duhamel's theorem (Carslaw and Jaeger, 1959). Although the two solutions are based on different approaches, they are equivalent in form. These solutions can be used to gain considerable insight into the conditions for which the Theis recovery method is appropriate for the analysis of dst data.

Correa and Ramey (1987) used their solution to demonstrate that the Theis recovery analysis is appropriate when the time since shut-in is large compared to the duration of the flow period. Under those conditions, the conventional recovery analysis can be employed to estimate hydraulic conductivity with the following equation:

$$K_r = \frac{2.30 q_{av}}{4\pi B m} \tag{7.4a}$$

where: q_{av} = the average flow rate during the flow period ($= (\pi r_c^2 (1 - h_s))/t_s$), $[L^2/T]$;
h_s = normalized head at shut-in, [dimensionless];
t_s = time of shut-in, [T];
m = the slope of the normalized head vs. $\log(t/t')$ plot, [dimensionless];
t = total time since the start of the test, [T];
t' = time since shut-in, [T].

An alternate form of the equation can be obtained by plotting the ratio of the normalized head over q_{av} vs. the time term, and using the slope of that plot to estimate K_r:

$$K_r = \frac{2.30}{4\pi B m_q} \tag{7.4b}$$

where: m_q = the slope of the normalized head over q_{av} vs. $\log(t/t')$ plot, $[T/L^2]$.

In groundwater investigations, the analysis of drillstem-test data essentially consists of the following three steps:

1. The normalized deviation from static (or the ratio of the normalized deviation from static over q_{av}) is plotted vs. the logarithm of (t/t′);
2. The slope of the straight line to which this plot converges as the time term goes to one is calculated;
3. The radial component of hydraulic conductivity is estimated using Equations (7.4a) or (7.4b).

Note that this method for the analysis of dst data is based on the assumption that the equilibrium head (also termed background or static head) of the formation is known, as is often the case in shallow groundwater investigations. If the equilibrium head is not known, the actual head, and not the normalized deviation from static, should be plotted on the y axis. In this case, the straight line identified on the plot should be projected to a time ratio of one. The head at that time should be a reasonable approximation of the equilibrium head of the formation.

There are several issues of practical importance for the drillstem test. The most significant of these are the dependence of dst responses on the normalized shut-in head (h_s), on the dimensionless storage parameter of the open-hole phase, and on the contrast between the dimensionless storage parameters of the open-hole (α_1) and shut-in (α_2) phases. Figure 7.7 is a normalized head vs. dimensionless time (β) plot of dst responses generated using the analytical solution of Karasaki (1990) for the case of α_2 two orders of magnitude larger than α_1. This figure clearly shows that a drillstem test can be configured so that the test duration is considerably shorter than that of a conventional open-hole slug test (solid line on Figure 7.7). Figure 7.8 is a plot in the format of the Theis recovery method in which the simulated responses of Figure 7.7 are normalized by the dimensionless flow rate term ($Q_d = (1 - h_s)r_c^2)/(2K_rBt_s)$ used in Karasaki (1990). The theoretical straight line (slope of 1.15) that would be predicted from the recovery analysis for responses plotted in this format is shown as a solid line in the figure. The dst data are normalized by Q_d in this and subsequent figures so that plots of the theoretical responses will converge on a single straight line. If the responses were not normalized in this manner, each curve would converge on a different straight line, the slope of which would be a function of the flow rate term. As shown in Figure 7.8, the smaller the head at shut-in (h_s), the later the time at which the test data approach the theoretical relationship.

The dst responses in Figure 7.8 converge on the theoretical line predicted by the Theis recovery method, even though flow into the well is continually decreasing during the flow period. Streltsova (1988) discusses the large error that can be introduced into recovery analyses for pumping tests by decreases in flow rate during the period of pumping. The reason that such rate decreases during the flow period of a drillstem test have no impact on the relationships depicted in Figure 7.8 is that the situation represented by a drillstem test is mathematically analogous to head recovery following a period of constant pumpage. The linear relationship found at late time in dst data is a product of the same set of assumptions that are employed in the Ferris-Knowles method (Ferris and Knowles, 1963), and it can be readily shown that the Ferris-Knowles assumptions are the slug-test equivalent of the Cooper-Jacob approximation (e.g.,

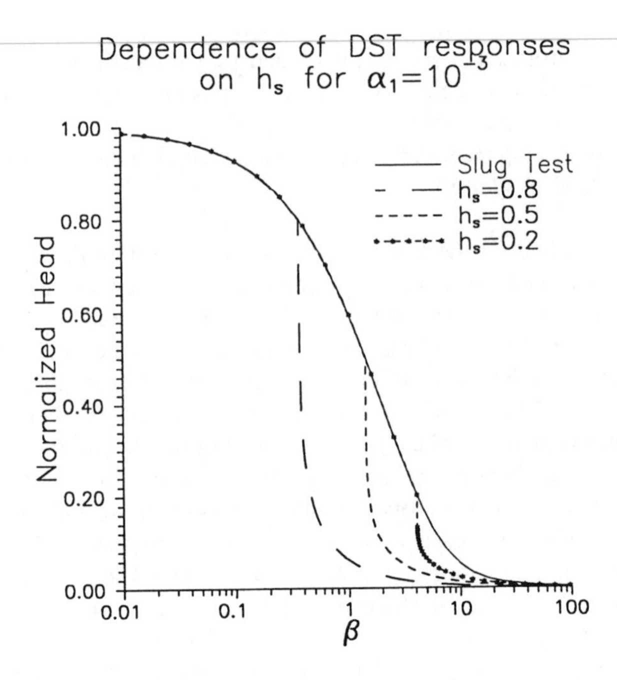

FIGURE 7.7 Normalized head ($H(t)/H_0$) vs. logarithm of β plots generated using the solution of Karasaki (1990). Note that normalized responses for a conventional open-hole slug test are included for comparison with the drillstem tests ($\alpha_1 = 0.01\alpha_2$; h_s and α_i defined in text).

Correa and Ramey, 1987; Peres et al., 1989). The major advantage of a drillstem test over a conventional slug test analyzed with the Ferris-Knowles method is that the well is shut-in at a relatively large normalized head, resulting in the Ferris-Knowles assumptions being applicable at a much larger normalized head than in a conventional slug test. Although it is not clear from Figure 7.8 because of the form of the ratio plotted on the y axis, the normalized head at which the simulated responses converge on a straight line is 0.11 and 0.07 for h_s equal 0.8 and 0.5, respectively; values that are much larger than the less than 0.0025 criterion recommended for the Ferris-Knowles method (Cooper et al., 1967).

Figure 7.8 displays relationships for a well-formation configuration in which $\alpha_1 = 0.001$. Figure 7.9 shows how these relationships depend on the magnitude of α_1 for the case of h_s equal to 0.5. This figure indicates that the normalized head at which a straight line relationship is obtained decreases with increases in α_1. Thus, if one suspects that α_1 could be quite large, the magnitude of H_0 should be increased to obtain better quality data at small normalized heads. Fortunately, however, the vast majority of well-formation configurations faced in shallow groundwater investigations will have an α_1 considerably less than 0.1.

The previous figures depict conditions for a two order of magnitude contrast between α_1 and α_2, a contrast that could readily be obtained if the open-hole portion of the test was performed in a well greater than or equal to 0.05 m in radius. Given the magnitude of equipment compressibility discussed in the previous section, such

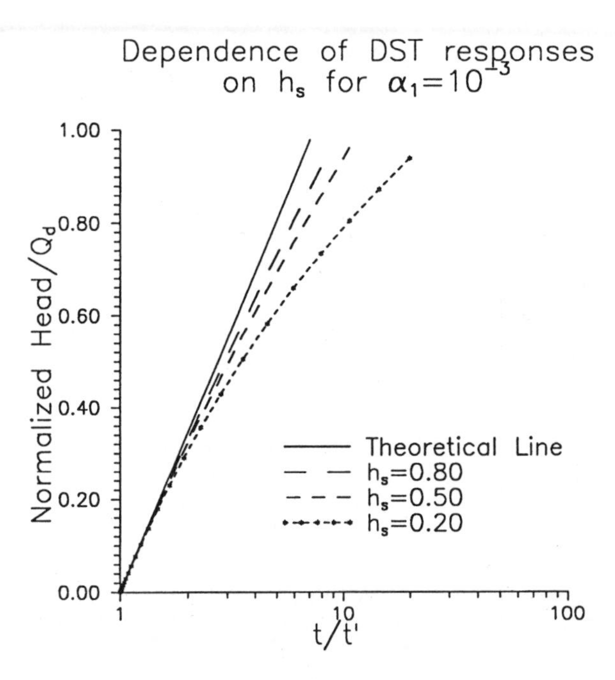

FIGURE 7.8 Theis recovery plot for hypothetical drillstem tests of Figure 7.7 (Q_d and t′ defined in text; theoretical line has slope of 1.15).

a contrast may not be obtainable in wells of smaller radii. Figure 7.10 illustrates how the magnitude of the α contrast impacts the convergence of dst data on the theoretical linear relationship for the case of h_s equal to 0.5. As shown in this figure, an α_1/α_2 ratio of 0.2 or less is needed to discern a linear relationship at normalized heads that are commonly measured in shallow groundwater investigations. At smaller contrasts, the normalized head at which the linear relationship is obtained approaches the 0.0025 criterion recommended for the Ferris-Knowles method.

Equations (7.4a) and (7.4b) and the discussion of Figures 7.8 to 7.10 are based on the assumption that the test is performed in a well that fully penetrates the formation. Often, however, drillstem tests are performed in wells that are screened over a limited vertical interval of the formation. Butler and Healey (1998) show how the Theis recovery analysis can be used to estimate hydraulic conductivity in wells that partially penetrate a formation. Thus, Equations (7.4a) and (7.4b) will also be appropriate for the analysis of drillstem tests in partially penetrating wells. However, the normalized head at which the dst responses will converge on the theoretical straight line may be much smaller than that shown in Figures 7.8 to 7.10. The viability of the approach for partially penetrating wells, i.e., the normalized head at which an approximate straight line can be attained, will primarily depend on the thickness of the formation and the magnitude of the vertical anisotropy ratio (K_z/K_r).

There are two additional issues of practical importance for drillstem tests. First, the technique is only appropriate for testing formations of moderately low to low hydraulic conductivity. In more permeable formations, the head response following

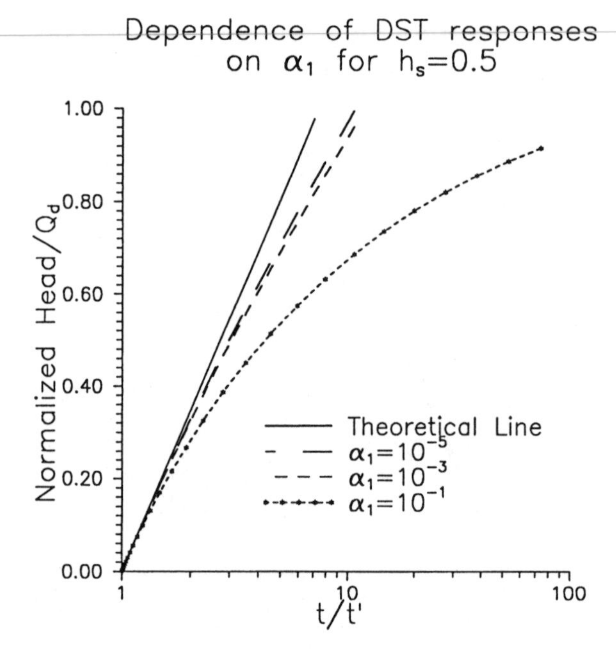

FIGURE 7.9 Theis recovery plot displaying the dependence of simulated responses on α_1 ($\alpha_1 = 0.01\alpha_2$; Q_d and t′ defined in text; theoretical line has slope of 1.15).

shut-in can be so rapid that it may be impossible to obtain sufficient noise-free data for analysis purposes. Second, although a drillstem test is often considerably shorter than a conventional open-hole slug test, a much larger volume of the formation is affected than in a shut-in slug test. Karasaki (1990) discusses how the volume of the formation being tested is a function of the normalized shut-in head. Given that discussion and the theoretical results illustrated in Figures 7.8 to 7.10, most drillstem tests should be performed using h_s between 0.9 and 0.5. In low-conductivity formations, h_s should be at the upper end of the range to minimize test duration while still impacting a reasonably sized volume of the formation.

Slug Tests in Formations of Extremely Low Hydraulic Conductivity

The discussion in the preceding sections was based on the assumption that conventional mathematical models of slug-induced flow (e.g., Equations [5.1a] to [5.1f]) are appropriate representations of the governing physics for tests in low-conductivity formations. Although that is often the case for the range of hydraulic conductivities faced in shallow groundwater investigations, it may not be true for studies associated with the siting of proposed hazardous waste repositories. In those situations, the permeability of the test interval may be so low that the influence of additional mechanisms, not accounted for in conventional models, may become significant.

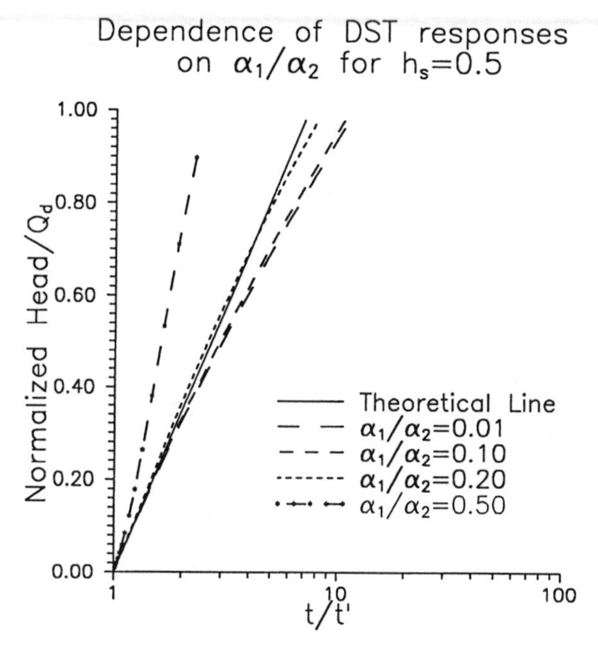

FIGURE 7.10 Theis recovery plot displaying the dependence of simulated responses on the ratio of α_1/α_2 (Q_d and t' defined in text; theoretical line has slope of 1.15).

Manifestations of these mechanisms include nonlinear compressibility of test equipment (r_c' varies with pressure), effects related to temperature disequilibrium (primarily produced by differences in the temperature of drilling fluids and formation waters), and incomplete recovery from pretest (e.g., drilling and development) pressure disturbances. Pickens et al. (1987), Beauheim (1994), and Novakowski and Bickerton (1997) provide further details concerning these mechanisms and recommend approaches to limit their impact on hydraulic conductivity estimates. Fortunately, however, such mechanisms have little effect on the vast majority of slug tests performed as part of shallow groundwater investigations.

8 The Analysis of Slug Tests — High Conductivity Formations

CHAPTER OVERVIEW

Conventional approaches for the analysis of response data from slug tests are based on variants of the mathematical model defined in Equations (5.1a) to (5.1f). That model, however, may not be an appropriate representation of the governing physics for slug tests performed in high-conductivity formations. Thus, conventional analysis approaches may introduce error into parameter estimates when applied in this hydrogeologic setting. In order to reduce the potential for such error, specialized techniques have been developed for the analysis of response data from tests performed in high-conductivity formations. The most common of these techniques are described in this chapter.

SLUG TESTS IN HIGH-CONDUCTIVITY FORMATIONS

In wells in formations of very high hydraulic conductivity or wells with very long columns of water above the top of the screen in formations of moderate or higher conductivity, it is not uncommon to observe slug-test response data that are oscillatory in nature. Figure 8.1 is an example of oscillatory response data from a slug test performed in a shallow well screened in the highly permeable sand and gravel deposits of the Kansas River alluvium in Douglas County, Kansas. Springer and Gelhar (1991) and McGuire and Zlotnik (1995) also report oscillatory responses from slug tests in similar hydrogeologic settings. Van der Kamp (1976) presents oscillatory data from tests in highly permeable sandstone aquifers, while Ross (1985) discusses oscillatory responses from permeable intervals in basalt flows located more than 1000 m below the land surface. In a less conventional application, Stone and Clarke (1993) report oscillatory response data from slug tests performed in ice-encased boreholes drilled into flow systems beneath active glaciers.

Although the oscillatory responses displayed in Figure 8.1 might appear quite distinct from the test data presented in previous chapters, oscillating water levels are actually just one end member of the spectrum of possible responses that can be observed in slug tests. Van der Kamp (1976), following the approach of Bredehoeft et al. (1966), adopts the terminology of classical physics to compare the range of possible responses to the behavior of a damped spring. According to this approach, oscillatory responses, such as those of Figure 8.1, are classified as underdamped and are characterized by a temporal behavior that is quite similar to a damped sinusoidal fluctuation. At the other end of the spectrum are nonoscillating responses, such as

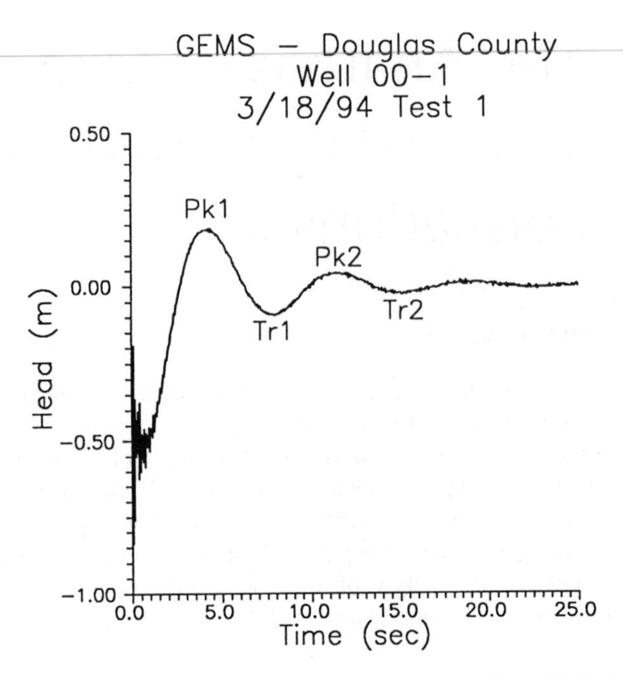

FIGURE 8.1 Head (H(t)) vs. time plot of a slug test performed in well 00-1 at the Geohydrologic Experimental and Monitoring Site (GEMS) in Douglas County, Kansas (Pki and Tri defined in text).

Figure 8.2 and the test data discussed in all previous chapters, which are classified as overdamped and are characterized by a temporal behavior that is quite similar to an exponential decay. Responses in the transition region between overdamped and underdamped are classified as critically damped. Figure 8.3 is an example of response data from this critically damped transition region. The rather subtle differences between responses in the overdamped (Figure 8.2) and critically damped (Figure 8.3) regions may often be obscured by background noise when plotted in the form of a linear head vs. time plot; so, a log head vs. time plot is more useful for identifying test data lying in the critically damped region. Data in the critically damped region will, as shown in Figure 8.4, display a distinct concave-downward curvature when plotted in a log head vs. time format, unlike response data from the overdamped region which display a near-linear to concave-upward character when plotted in the same format (e.g., Figure 8.5 and many of the figures of earlier chapters). However, as discussed in Chapter 2, response data from tests in wells screened across the water table under certain conditions can also display a concave-downward form when plotted in a log head vs. time format. Thus, some consideration must be given to the specifics of the well-formation configuration before classifying response data with a concave-downward curvature as critically damped.

The wide range of behaviors depicted in Figures 8.1 to 8.5 is the result of the influence of additional mechanisms that are not accounted for in conventional models of slug-induced responses (e.g., Equations [5.1a] to [5.1f]). Although several factors are responsible for the character of the responses observed in the underdamped and

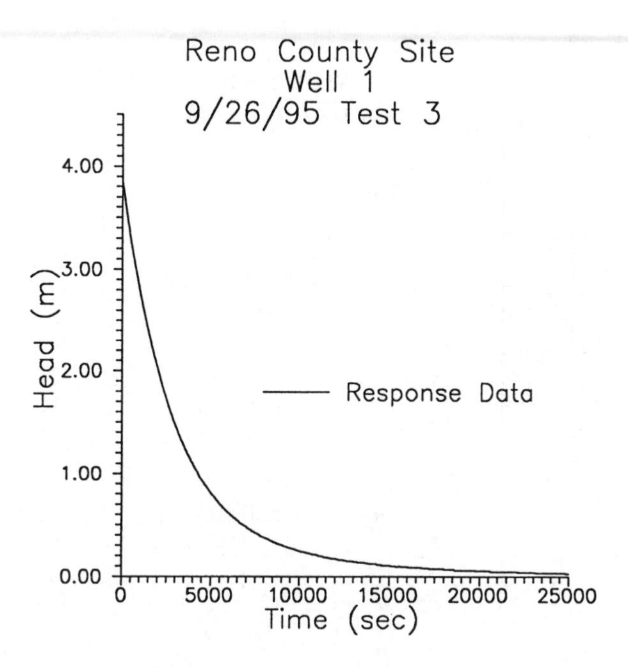

FIGURE 8.2 Head (H(t)) vs. time plot of a slug test performed in well 1 at a monitoring site in Reno County, Kansas.

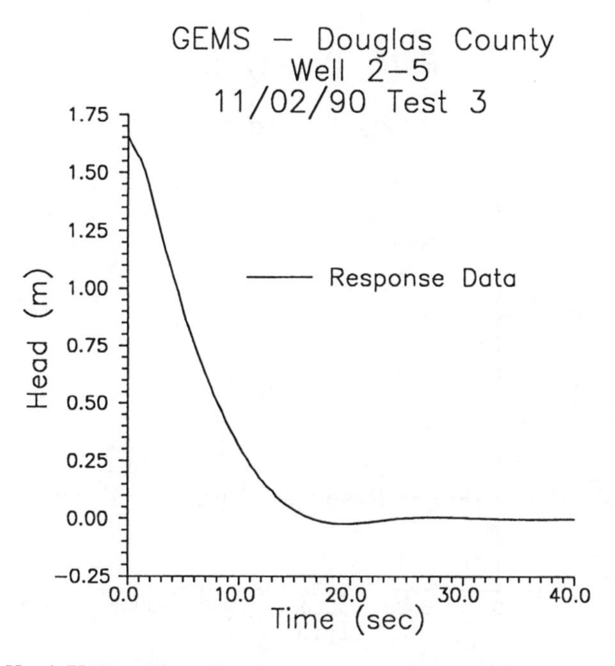

FIGUR 8.3 Head (H(t)) vs. time plot of a slug test performed in well 2-5 at the Geohydrologic Experimental and Monitoring Site (GEMS) in Douglas County, Kansas.

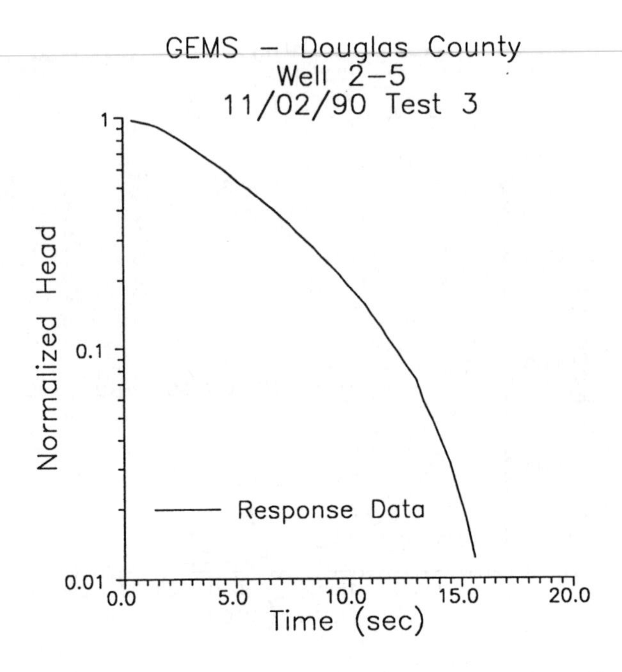

FIGURE 8.4 Logarithm of normalized head ($H(t)/H_0$) vs. time plot of a slug test performed in well 2-5 at the Geohydrologic Experimental and Monitoring Site (GEMS) in Douglas County, Kansas.

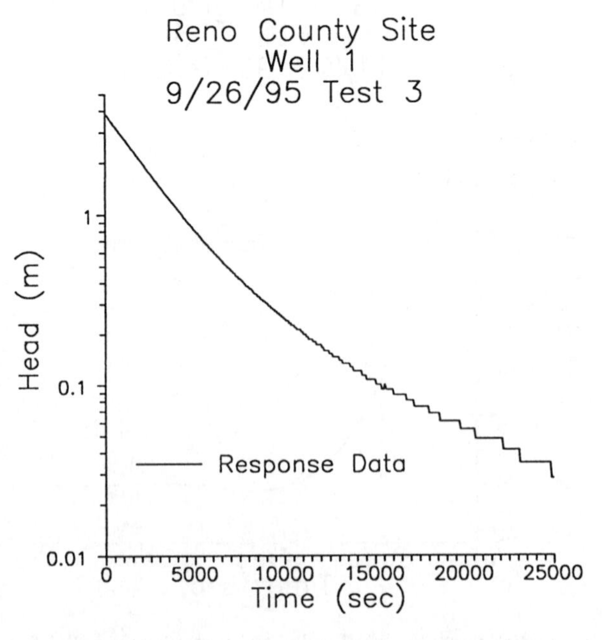

FIGURE 8.5 Logarithm of head ($H(t)$) vs. time plot of a slug test performed in well 1 at a monitoring site in Reno County, Kansas.

critically damped regimes, the primary controlling mechanism is the inertia of the water column in the well (Bredehoeft et al., 1966). In the underdamped regime, the momentum of the water column is great enough that more water flows into or out of the well than would be predicted from the conventional overdamped model (e.g., Cooper et al., 1967). This additional flow results in a head difference that produces the oscillatory character of the responses. Bredehoeft et al. (1966) show that the transmissivity of the screened interval and the length of the water column above the top of the screen are the primary determinants of the significance of the momentum of the water column. In certain conditions, additional frictional losses that are not considered in the conventional overdamped model can also be important controls on slug-induced responses (e.g., Ross, 1985; McElwee and Butler, 1996).

Several mathematical models have been proposed to represent the primary physical mechanisms affecting slug-induced responses in high-conductivity formations (e.g., Van der Kamp, 1976; Krauss, 1977; Kipp, 1985; Springer and Gelhar, 1991; McElwee and Zenner, 1993). In their most general form, these models, all of which have been derived from conservation of momentum considerations (Bird et al., 1960), can be written in terms of the water level in the test well as:

$$\frac{(L_e + w)}{g}\frac{d^2w}{dt^2} + \frac{A}{g}\left(\frac{dw}{dt}\right)^2 + \frac{F}{g}\frac{dw}{dt} + w = h \tag{8.1a}$$

$$w(0) = H_0 \tag{8.1b}$$

$$\frac{dw}{dt} = 0, \; t = 0 \tag{8.1c}$$

where: L_e = effective length of the water column, [L];
w = deviation of water level from static level in the test well, [L];
A = parameter for nonlinear term, [dimensionless];
F = viscous loss parameter, [L/T];
h = deviation of head in the formation from static conditions, [L].

The models differ in the assumptions that they adopt concerning the (L_e+w) and squared temporal derivative (squared velocity) terms, and in their representation of slug-induced flow in the formation. These differences will be briefly summarized in the following paragraphs.

The initial work on slug tests in high-conductivity formations assumed that the effective length of the water column is equal to the length of the column above the screen plus three eighths of the screen length, a quantity that resulted from the derivation of Bredehoeft et al. (1966). Kipp (1985) proposed a modified definition of the effective column length (the length of the column above the screen plus one half of the screen length for wells in which the screen and casing radii are equal), while McGuire and Zlotnik (1995) proposed a further modification that included changes in casing radius associated with a straddle-packer test system. However, both Kipp (1985) and Kabala et al. (1985) independently noted that for the analysis of field data the effective column length often had to be considerably larger than a definition based on geometrical considerations. McElwee and Zenner (1993) have shown that an effective column length larger than that which would be predicted from geometrical considerations is a result of additional kinetic energy contributions and energy losses associated with changes in the casing radius. They recommend that the effective column length be a parameter that is estimated as part of the analysis, a procedure that had previously been adopted on pragmatic grounds by several investigators (e.g., Kabala et al., 1985). Note that most authors have assumed that $L_e \gg w$ and have ignored the nonlinearity introduced by the w in the (L_e+w) term. Kabala et al. (1985) show that the effect of this nonlinearity is quite small as long as H_0 is less than 25% of the effective column length.

Except for the model of McElwee et al. (McElwee and Zenner, 1993; McElwee and Butler, 1996), the squared velocity term in Equation (8.1a) has been effectively ignored in models of slug tests in high-conductivity formations. Kabala et al. (1985) present a numerical evaluation of the significance of this term for the case of A equal to 0.75, a value resulting from the theoretical derivation of Bredehoeft et al. (1966). Their results indicate that the squared velocity term has virtually no impact on slug-induced responses, a conclusion that was also reached by McGuire and Zlotnik (1995) in the context of an investigation of multilevel slug tests. However, field data collected from tests in a highly permeable sand and gravel aquifer (e.g., McElwee and Butler, 1996) indicate that the magnitude of the A parameter in Equation (8.1a) can be much larger than that assumed by Kabala et al. (1985) or McGuire and Zlotnik (1995).

The h term in Equation (8.1a) links the water in the well to that in the formation. Therefore, in order to have a complete mathematical representation of a slug test, a model must be adopted to represent flow in the formation. Several authors (e.g., Van der Kamp, 1976; Krauss, 1977; Kipp, 1985; Kabala et al., 1985) have employed the conventional radial flow equation (Equation [5.1a]) to represent the slug-induced flow. In this case, however, the resulting solution is in a rather complicated form that may not be amenable to ready analysis. Other authors have obtained a simpler form for the solution by exploiting the insensitivity of slug-test responses to the dimensionless storage parameter (α), and adopting a quasi-steady-state representation of flow in the formation. Springer and Gelhar (1991) employ the Bouwer and Rice model of slug-induced flow, while McElwee et al. (McElwee et al., 1992; McElwee and Zenner, 1993) use the Hvorslev model.

The preceding overview indicates that a considerable amount of work has been done on slug tests in high-conductivity formations. This body of work has essentially resulted in three classes of methods for the analysis of response data. Representative techniques from each of these classes are described in the following sections.

The Van der Kamp Model

The most commonly used method for the analysis of response data from slug tests in fully penetrating wells in high-conductivity formations is based on a model that was originally developed by Van der Kamp (1976). This approach involves simplifying Equation (8.1a) to the following form:

$$\frac{d^2w_d}{dt_d^2} + C_d\frac{dw_d}{dt_d} + w_d = 0,\ C_d \ll 2 \tag{8.2}$$

where: w_d = w/H_0, [dimensionless];
t_d = $(g/L_e)^{0.5}t$, [dimensionless];

$$C_d = \sqrt{\frac{g}{L_e}}\frac{r_c^2\ln\left[\left(1.27/r_w^2\right)\left(L_e/g\right)^{0.5}\left(K_r/S_s\right)\right]}{8K_rB}, \text{ [dimensionless].}$$

The key feature of Equation (8.2) is the qualifier that C_d, the dimensionless damping parameter, must be much less than two, i.e., the solution is only appropriate for responses that lie well within the underdamped region. In the definition for C_d used in Equation (8.2), a value of two corresponds to the point of critical damping. Note that some authors define the damping parameter as $C_d/2$; so, a value of one is the point of critical damping under that definition.

The solution of Equation (8.2) is the damped spring solution of classical physics (Kreyszig, 1979) and can be written for the initial conditions given in Equations (8.1b) and (8.1c) as:

$$w_d\left(t_d\right) = e^{-(C_d/2)t_d}\cos\left(\omega_d t_d\right) \tag{8.3}$$

where: ω_d = frequency parameter (= $(1-(C_d/2)^2)^{0.5}$), [dimensionless].

Assuming that Equation (8.3) is an appropriate model of test responses in the underdamped region, damping (C) and frequency (ω) parameters can be estimated from the response data. These estimates can then be used to calculate the effective column length and the dimensionless damping parameter:

$$L_e = \frac{g}{\left(\omega^2 + (C/2)^2\right)} \tag{8.4a}$$

$$C_d = \frac{C}{\sqrt{g/L_e}} \tag{8.4b}$$

Van der Kamp (1976) proposes an iterative approach for estimation of hydraulic conductivity using the effective column length and the dimensionless damping parameter in the following equation:

$$K_{r(n)} = m_2 + m_1\ln\left(K_{r(n-1)}\right) \tag{8.5}$$

where: $K_{r(n)}$ = hydraulic conductivity estimate for iteration n, [L/T];
$m_1 = (r_c^2/B)(g/L_e)^{0.5}/4C_d$;
$m_2 = -m_1 \ln[0.79 r_w^2 S_s (g/L_e)^{0.5}]$.

The initial estimate for K_r should be set to the m_2 parameter of Equation (8.5).

The Van der Kamp method essentially consists of the following four steps:

1. The response data are plotted vs. the time since test initiation;
2. The damping and frequency parameters are estimated from subsequent peaks or troughs in the test data using the following equations:

$$C = \frac{2\ln(W_n/W_{n+1})}{t_{n+1} - t_n} \tag{8.6a}$$

$$\omega = \frac{2\pi}{t_{n+1} - t_n} \tag{8.6b}$$

where: W_n = the w value at the nth peak or trough in the test data, [L];
t_n = time of the nth peak or trough in the test data, [T];

3. L_e and C_d are calculated from the damping and frequency parameters using Equations (8.4a) to (8.4b);
4. Hydraulic conductivity is estimated with Equation (8.5) or an analogous nomograph procedure suggested by Uffink (Kruseman and de Ridder, 1990).

Semi-automated analogs of these procedures have been proposed by Sepulveda (1992) and Wylie and Magnuson (1995).

An example can be used to demonstrate a field application of the Van der Kamp method. In March of 1994, a series of slug tests were performed at a monitoring well at a research site in the Kansas River alluvium in Douglas County, Kansas. The well was screened in a sequence of coarse sand and gravel. Table 8.1 summarizes the pertinent well-construction information, while Table 8.2 presents a subset of the test data. Figure 8.1 is a plot of the response data, Pki and Tri are the ith peak and trough, respectively, of the data record. Table 8.3 summarizes the results of the Van der Kamp analysis using the peaks (case 1) and troughs (case 2) of the data record. Although there is some uncertainty about the timing of the second peak and trough, the resulting hydraulic conductivity estimates are within 40% of one another. The C_d estimates given in Table 8.3 indicate that the response data lie well within the underdamped region, as is required for the Van der Kamp method.

There are several issues of practical importance with respect to the Van der Kamp method. First, the Van der Kamp method assumes that the test well is fully screened across a confined aquifer. Thus, when this method is employed to analyze response data from tests performed in partially penetrating wells, it can lead to an

TABLE 8.1
Well Construction Information for Well 00-1 at the Geohydrologic Experimental and Monitoring Site

Well Designation	r_w(m)	r_c(m)	b(m)	B(m)	d(m)
Douglas County GEMS Well 00-1	0.025	0.025	0.76	10.67	5.61

TABLE 8.2
Response Data from 3/18/94 Test #1 in Well 00-1 at the Geohydrologic Experimental and Monitoring Site

Time (s)	Head (m)
0.25	–0.552
0.35	–0.402
0.45	–0.375
0.55	–0.476
0.65	–0.570
0.75	–0.585
0.85	–0.518
0.95	–0.454
1.05	–0.457
1.15	–0.488
1.20	–0.415
1.30	–0.415
1.45	–0.372
1.50	–0.378
1.60	–0.326
1.70	–0.314
1.80	–0.293
1.90	–0.235
2.00	–0.207
2.15	–0.155
2.25	–0.131
2.40	–0.091
2.55	–0.040
2.70	0.003
2.85	0.037
3.00	0.076
3.20	0.107
3.35	0.140
3.55	0.159
3.80	0.177

TABLE 8.2 (continued)
Response Data from 3/18/94 Test #1 in Well 00-1 at the Geohydrologic Experimental and Monitoring Site

Time (s)	Head (m)
4.00	0.183
4.25	0.180
4.50	0.183
4.75	0.168
5.05	0.143
5.35	0.113
5.65	0.079
6.00	0.034
6.35	–0.009
6.70	–0.037
7.10	–0.064
7.50	–0.088
7.95	–0.095
8.45	–0.082
8.95	–0.052
9.45	–0.027
10.05	0.003
10.60	0.021
11.25	0.040
11.90	0.037
12.60	0.024
13.35	0.003
14.15	–0.018
15.00	–0.021
15.85	–0.018
16.80	–0.006
17.80	0.006
18.85	0.012
20.00	0.006
21.15	0.003
22.40	0.000
23.75	–0.003

overestimation of the hydraulic conductivity of the aquifer. In the case of the above example, slug-induced vertical flow could easily be the source of an overestimation in hydraulic conductivity on the order of a factor of two (e.g., Figure 5.12). Thus, the Van der Kamp method must be used quite cautiously with partially penetrating wells of small to moderate aspect ratios.

Second, Equation (8.5) requires estimates for the specific storage of the formation and the effective radius of the well screen, parameters for which there may be considerable uncertainty. However, these two quantities only appear as a product

TABLE 8.3
Results of Analysis of GEMS 00-1 3/18/94 Test #1 Response Data with Van der Kamp Method

	$C(s^{-1})$	$\omega(s^{-1})$	$L_e(m)$	C_d	$K_r(m/d)$
Case 1	0.422	0.891	11.70	0.461	449.
Case 2	0.332	0.867	12.59	0.376	618.

Note: Row and column labels defined in text.

inside the logarithmic term of m_2; so, even a relatively large error in their magnitude will have a relatively small effect on the conductivity estimate. Uffink (Kruseman and de Ridder, 1990) proposes an extension of the Van der Kamp method to the case of an infinitely thin well skin. In this approach, the well skin is included in the analysis by adjusting the effective screen radius (r_w) of Equation (8.5) as suggested by Ramey et al. (1975) and discussed further in Chapter 9.

A third issue of practical importance is that of the magnitude of H_0. In all of the examples presented by Van der Kamp (1976), H_0 is on the order of a few centimeters or less. Van der Kamp (1976; pers. commun., 1996) emphasizes that H_0 must be kept quite small for Equation (8.2) to be an appropriate representation of the governing physics. For similar reasons, Van der Kamp (1976) recommends that the early-time data be ignored, and that only data after the first peak or trough in the record be used in the analysis. Given the small value of H_0 recommended for this method and the emphasis on the later portions of the data record, it is imperative that sensor and other background noise be quite small.

Fourth, as discussed in Chapter 3, there may be considerable uncertainty about the value of H_0 for tests in high-conductivity formations. An advantage of the Van der Kamp method is that an estimate of H_0 is not actually required for the analysis. Equations (8.6a) and (8.6b) demonstrate that the focus of the analysis approach is on the period of the oscillations and the difference between subsequent peaks or troughs in the record, and not the actual head values.

Finally, Van der Kamp (1976) provides a brief analysis to show that frictional (viscous) losses within the well casing will be small under conditions expected in most shallow groundwater investigations. The exceptions to this are tests performed in wells with very small casing radii and tests in which water-level fluctuations are very slowly damped. Ross (1985) points out that this second condition is often found in tests performed in very deep wells screened in intervals of moderate to high hydraulic conductivity. He proposes a modification of the Van der Kamp method that includes the addition of a term to account for frictional losses within the well casing.

General Linear Methods

The Van der Kamp method is only appropriate for response data that lie well within the underdamped region (C_d of Equation (8.2) much less than two). In order to have a more general procedure for the analysis of response data affected by the momentum of the water column, several authors have proposed methods that can be used over the entire range of underdamped through overdamped responses. Three of these methods have been fairly extensively evaluated in the field (Kipp, 1985; Springer and Gelhar, 1991; McElwee et al., 1992) and will be the focus of this discussion.

All of the approaches discussed in this section involve simplifying Equation (8.1a) to the following form:

$$\frac{d^2 w_d}{dt_d^2} + w_d = h_d \tag{8.7}$$

where: $h_d = h/H_0$, [dimensionless].

These methods primarily differ in the manner that flow in the formation is represented in the underlying mathematical model. Despite the different conceptualizations of the slug-induced flow, however, all of the proposed field methods are based on solutions to the following equation:

$$\frac{d^2 w_d}{dt_d^2} + C_d \frac{dw_d}{dt_d} + w_d = 0 \tag{8.8}$$

where: $C_d = \sqrt{\frac{g}{L_e}} \frac{r_c^2 \ln\left[(L_e/g)^{0.5}(K_r/S_s r_w^2)\right]}{4K_r B}$, the Kipp (1985) model;

$= \sqrt{\frac{g}{L_e}} \frac{r_c^2 \ln[R_e/r_w]}{2K_r b}$, the Springer and Gelhar (1991) model;

$= \sqrt{\frac{g}{L_e}} \frac{r_c^2 \ln\left[1/(2\psi) + \left(1 + (1/(2\psi))^2\right)^{0.5}\right]}{2K_r b}$, the linearized variant of the McElwee et al. (1992) model.

In all cases, the methods involve estimating L_e and C_d from the response data, and then calculating hydraulic conductivity from the definition of C_d. Each of these methods is briefly summarized in the following paragraphs.

Kipp (1985) was apparently the first to propose an analysis method that can be used over the entire range of possible responses. Flow in the formation is represented using the standard radial-flow equation (Equation [5.1a]), and a semianalytical solution to Equation (8.7) is obtained that is valid over the entire range from underdamped to overdamped conditions. However, the complexity of the solution makes it of limited use for the analysis of field data. In order to develop a practical field method, Kipp proposes an approximate type-curve approach based on the solution to Equation (8.8). This method is founded on two assumptions concerning the well-aquifer configuration: (1) the magnitude of the dimensionless storage parameter (α) is small; and (2) the magnitude of the parameter group $((L_e/g)\ (K_r/S_s r_w^2)^2)$ is very large. Both of these assumptions should be quite reasonable for tests in formations of moderate to high hydraulic conductivity. As with the Van der Kamp method, an estimate of specific storage is necessary for the calculation of hydraulic conductivity. Although the solution to Equation (8.7) is of little use for practical applications, Kipp (1985) employs this solution to demonstrate that the approximations incorporated in the Van der Kamp method are appropriate for C_d values less than about 0.4 and that the momentum of the water column can be ignored for C_d values greater than about 10.

The major limitation of the Kipp method is that it only applies to response data from wells that are fully screened across a confined formation. The Springer and Gelhar method and the linearized variant of the McElwee et al. method have been developed for use in partially penetrating wells. As shown in Equation (8.8), the Springer and Gelhar method adopts the Bouwer and Rice (1976) model of flow in an unconfined formation, while the McElwee et al. method uses the Hvorslev (1951) model of flow in a confined formation. Thus, these two methods are based on a quasi-steady-state representation of the slug-induced flow.

For all three methods, the solution of Equation (8.8) can be presented in the form of type curves that are a function of C_d. Figure 8.6 displays a series of these curves for selected values of C_d. A C_d value can be estimated by matching the test data to the type curves, either via manual curve matching or an automated analog. Hydraulic conductivity can then be calculated using the definition of C_d given in Equation (8.8). Note that McElwee and Zenner (1993) extend the method to include viscous losses in the well casing following the approach of Ross (1985). This involves the addition of a $(1 + (8\nu L/r_c^2 g))$ term to the C_d definition of Equation (8.8), where ν is the kinematic viscosity $[L^2/T]$ and L is the nominal length of the water column [L]. McGuire and Zlotnik (1995) extend the Springer and Gelhar method to the case of multilevel slug tests performed with straddle packers. They show that the nature of the responses is heavily dependent on the geometry of test equipment (e.g., dimensions of the straddle-packer system and associated piping) as a result of the impact of those factors on L_e.

When attempting to perform slug tests in high-conductivity formations, it can be extremely difficult to introduce the slug in a manner that is near-instantaneous relative to the formation response. Noninstantaneous slug introduction can make it very difficult to estimate exactly when a test begins and what value should be used for H_0. Since both the Springer and Gelhar, and McElwee et al. models employ a quasi-steady-state representation of the slug-induced flow, these models do not strictly require that the slug be introduced in a near-instantaneous fashion. Thus, the translation method of Pandit and Miner (1986) can be extended to tests in high-conductivity formations when these models are a reasonable representation of the governing physics. However, in the case of underdamped or critically damped responses, the translation method must be modified to honor the condition that the initial velocity of the water column is zero (Equation (8.1c)). Unlike the overdamped case, an arbitrary time cannot be used as the time of test initiation. Instead, only times at which peaks or troughs appear in the data record (i.e., points at which the velocity of the water column is zero) can be used as the time of test initiation. Data

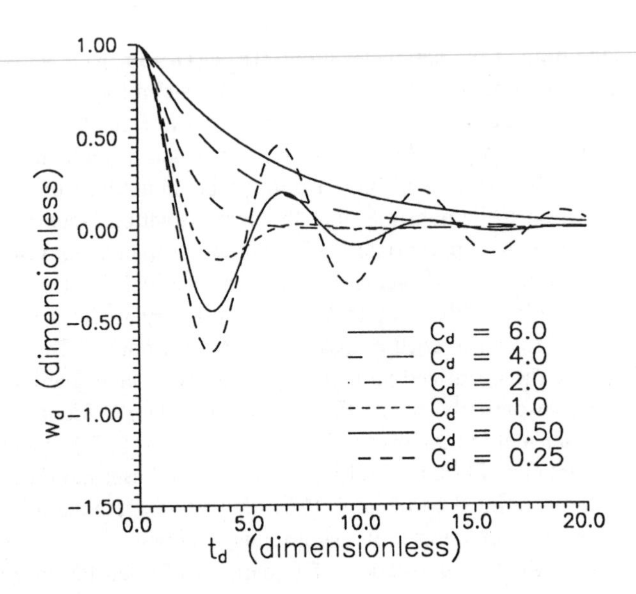

FIGURE 8.6 Normalized deviation from static (w/H_0) vs. dimensionless time type curves generated using any one of the three general linear methods (t_d and C_d defined in text).

prior to the chosen peak or trough are truncated from the data record, and test times are recalculated with respect to the peak or trough that has been selected as the assumed starting point.

Data analysis with one of the general linear methods essentially consists of the following three steps:

1. The response data are plotted vs. the time since test initiation. If the data are oscillatory in nature and there is concern about the impact of noninstantaneous slug introduction, the time of test initiation used for the analysis should be the time of a peak or trough in the record. Data prior to that time should be ignored;
2. The data plot is overlain by a plot of the C_d type curves. A match point for t_d is determined without shifting either axis, and a value for C_d is interpolated from the plotted type curves. L_e is calculated from the definition of t_d given in Equation (8.2). For oscillatory data, a check on the L_e and C_d values can be obtained by estimating the damping and frequency parameters from subsequent peaks or troughs in the test data with Equations (8.6a) and (8.6b), and then using these estimates to calculate L_e and C_d with Equations (8.4a) and (8.4b);
3. Hydraulic conductivity is estimated by using the definition of C_d given in Equation (8.8) rewritten as follows:

$$K_r = \sqrt{\frac{g}{L_e}} \frac{r_c^2 \ln\left[\left(L_e/g\right)^{0.5}\left(K_r/S_s r_w^2\right)\right]}{4BC_d} - \text{Kipp} \qquad (8.9a)$$

$$= \sqrt{\frac{g}{L_e} \frac{r_c^2 \ln[R_e/r_w]}{2bC_d}} \text{ – Springer and Gelhar} \qquad (8.9b)$$

$$= \sqrt{\frac{g}{L_e} \frac{r_c^2 \ln\left[1/(2\psi) + \left(1 + (1/(2\psi))^2\right)^{0.5}\right]}{2bC_d}} \text{ – McElwee et al.} \qquad (8.9c)$$

Although manual curve matching is possible with these models, it can be quite involved. Thus, use of automated matching procedures (e.g., McGuire and Zlotnik, 1995) is strongly recommended.

The same slug test used to demonstrate the Van der Kamp method can be employed to illustrate analysis procedures with one of the generalized linear models, specifically, the linearized variant of the McElwee et al. method. Since this test has clearly been impacted by noninstantaneous slug introduction (Butler et al., 1996), the translation method must be utilized in the analysis. Figure 8.7 is a plot of the response data of Figure 8.1 in which data before the first peak (Pk1) have been truncated from the record, and test times have been recalculated with respect to Pk1. Figure 8.8 is a plot of the reinitialized data and the best-fit C_D type curve. A L_e value of 12.56 m is calculated from the t_d match point and a C_d value of 0.43 is estimated from the type curve match. Substitution of this L_e value and the geometric parameters of Table 8.1 into Equation (8.9c) yields a hydraulic conductivity estimate of 255 m/d. This estimate can be checked with the alternate method described in step two of the analysis procedure. Use of the L_e and C_d estimates given in Table 8.3 with Equation (8.9c) produces hydraulic conductivity estimates of 242 m/d and 283 m/d for the analyses based on the peaks (case 1) and troughs (case 2), respectively, values which are in good agreement with that obtained with the automated curve-matching procedure. Note that the average of the hydraulic conductivity estimates obtained with the Van der Kamp model is approximately a factor of two greater than that found using the linearized McElwee et al. model, yet a further demonstration of the significance of vertical flow in slug tests in partially penetrating wells.

The issue of most practical importance for this class of methods is the magnitude of H_0. If a moderate to large value of H_0 is used, Equations (8.7) and (8.8) may not be reasonable representations of the governing physics. Thus, as recommended in the previous section, H_0 should be kept as small as possible.

NONLINEAR MODELS

The analysis methods presented in the previous two sections are based on mathematical models in which it is assumed that the squared velocity term of Equation (8.1a) is negligible and that L_e is much larger than H_0 (i.e., the mathematical models are linear in w_d). If these linear models are accurate representations of the physical mechanisms occurring during a slug test, then the response data from a series of slug tests performed with different H_0 should coincide when plotted in a normalized head vs. time format. However, in formations of very high hydraulic conductivity, it is not

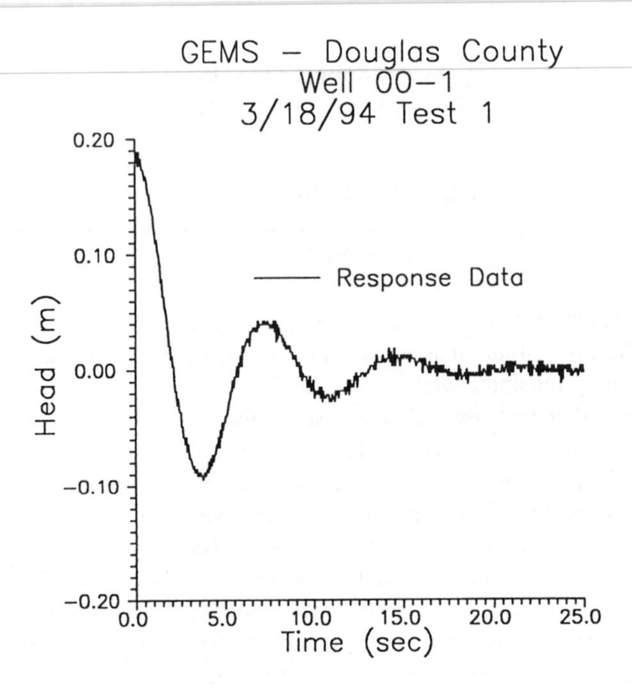

FIGURE 8.7 Head (H(t)) vs. time plot of slug test of Figure 8.1 after application of translation method (response data prior to Pk1 truncated from record and test times reinitialized with respect to Pk1).

uncommon to observe response data that display a reproducible dependence on H_0 (e.g., Butler et al., 1996), an indication that nonlinear mechanisms are having a significant effect on the slug-induced responses.

A review of a large number of slug tests that have been performed in formations of very high hydraulic conductivity (e.g., McElwee and Zenner, 1993; McGuire and Zlotnik, 1995; McElwee and Butler, 1996) reveals that the impact of nonlinear mechanisms will vary from well to well at a site. Figure 8.9 displays response data from a series of tests performed in a different well in the same highly permeable sand and gravel aquifer as in Figure 8.1. In this case, the response data lie in the critically damped (Tests 6 and 9) and slightly overdamped (Test 15) regimes. These data clearly demonstrate the need to keep H_0 as small as possible when performing slug tests in formations of high hydraulic conductivity. Even when an H_0 of a moderate size is used, such as Test 6 (H_0 = 1.05 m), hydraulic conductivity can be underestimated by a factor of 2 or more when conventional linear models are employed for the analysis. Although use of response data from tests initiated with very small H_0, such as Test 15, may yield hydraulic conductivity estimates that have not been significantly affected by nonlinear mechanisms, sensor and other background noise can introduce considerable uncertainty into those estimates.

The most striking examples of the influence of nonlinear mechanisms on slug tests in formations of high hydraulic conductivity come from wells in which the response data display effects similar to those shown in Figure 8.10. In the series of

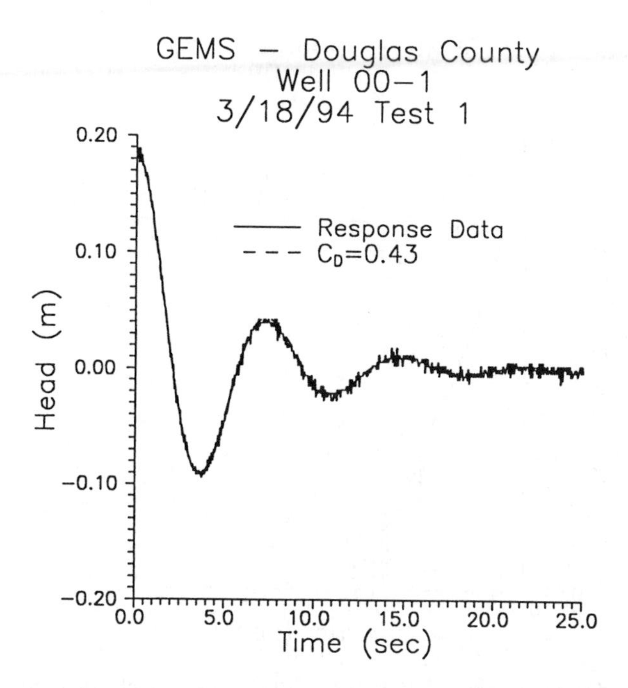

FIGURE 8.8 Head (H(t)) vs. time plot of slug test of Figure 8.1 after application of translation method and best-fit C_d type curve. Note that type curve is virtually indistinguishable from the plot of the response data.

tests displayed in that figure, the apparent regime of the response data varies from critically damped to underdamped as a function of H_0. Of particular significance is the similarity of the data displayed in Figure 8.10 to the type curves presented in Figure 8.6. In this case, use of one of the general linear models for the analysis of tests initiated with moderate to large H_0 can easily lead to an underestimation of hydraulic conductivity on the order of a factor of four.

When response data exhibit the characteristics displayed in Figures 8.9 and 8.10, a nonlinear model (e.g., Stone and Clarke, 1993; McElwee and Zenner, 1993) may be the most appropriate representation of the relevant physics. Although none of the analysis procedures based on proposed nonlinear models can yet be considered viable field methods, it is worthwhile to briefly review the general form of such models. The nonlinear model of McElwee et al. (McElwee and Zenner, 1993; McElwee and Butler, 1996) has received the most use in conventional hydrogeologic applications; so, it will be the focus of this brief discussion.

The variant of the McElwee et al. model considered here involves simplifying Equation (8.1a) and rewriting it in a dimensionless form as:

$$\frac{d^2w_d}{dt_d^2} + A_d\left(\frac{dw_d}{dt_d}\right)^2 + C_d(F_d+1)\frac{dw_d}{dt_d} + w_d = 0 \qquad (8.10)$$

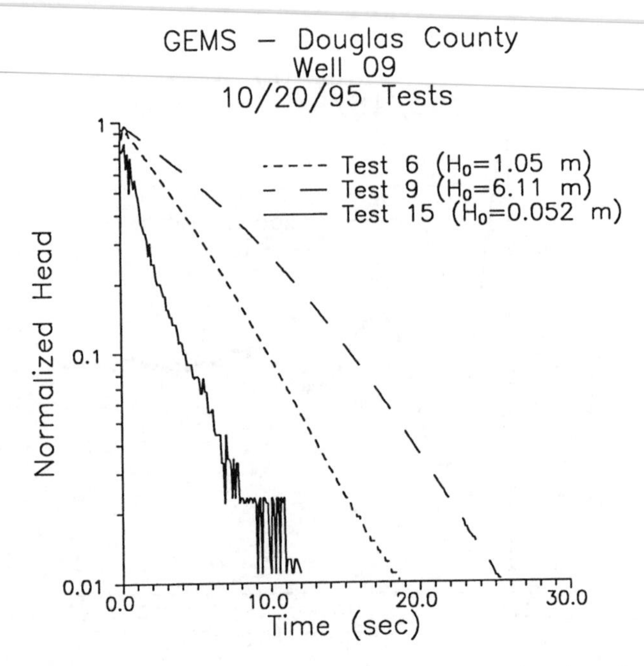

FIGURE 8.9 Logarithm of normalized head ($H(t)/H_0$) vs. time plot of a series of slug tests performed in well 09 at the Geohydrologic Experimental and Monitoring Site (GEMS) in Douglas County, Kansas.

where: A_d = $(AH_0)/L_e$, [dimensionless];
A = parameter of nonlinear term given in Equation (8.1a), [dimensionless];
C_d = McElwee et al. damping parameter of Equation (8.8), [dimensionless];
F_d = viscous loss parameter (= $(8\nu L)/(r_c^2 g)$), [dimensionless};
L = nominal length of the water column, [L];
ν = kinematic viscosity, [L^2/T].

The primary simplifications involved in going from Equation (8.1a) to Equation (8.10) are that the effective column length is much greater than H_0 and that the flow in the formation can be represented with the model of Hvorslev (1951). Note that the first assumption is not required by the McElwee et al. model in its most general form, but is invoked here because it is appropriate in the majority of field applications.

There are several issues of practical importance with respect to the field application of the McElwee et al. model. First, analysis methods based on this model require the estimation of three parameters (A, L_e, and C_d), so the estimation procedure can be more involved than with the techniques described in the two previous sections. Thus, this method should only be applied when nonlinear mechanisms are clearly having a significant impact on test responses. A program of slug tests at a well screened in highly permeable material can be designed to assess the significance of nonlinear mechanisms. As described in Chapter 2, this can be done by performing

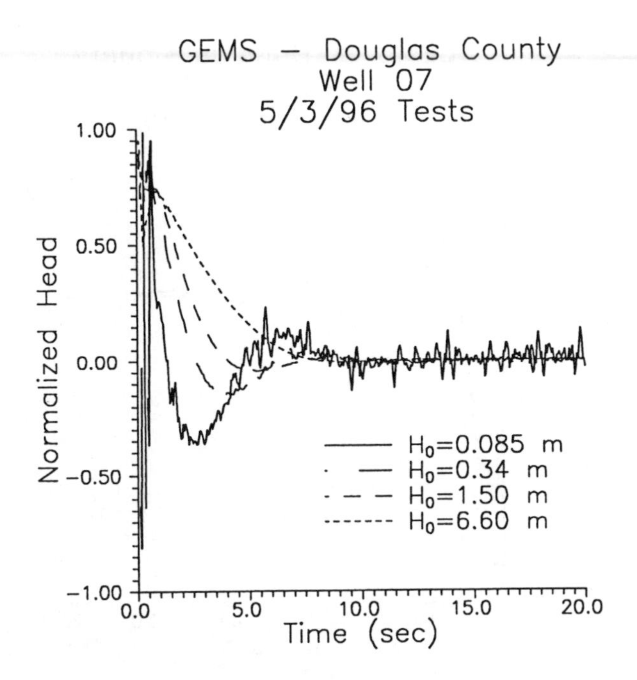

FIGURE 8.10 Logarithm of normalized head ($H(t)/H_0$) vs. time plot of a series of slug tests performed in well 07 at the Geohydrologic Experimental and Monitoring Site (GEMS) in Douglas County, Kansas. Note the increase in the relative magnitude of the noise introduced by the data acquisition equipment with smaller H_0.

repeat slug tests in which H_0 is varied between tests. In order to reliably assess the role of nonlinear mechanisms, the lower end of the H_0 range should be between 0.1 to 0.2 m, while the upper end should be above 1 m. If normalized plots of test responses over this H_0 range approximately coincide, nonlinear mechanisms can be ignored. Use of a narrow range of H_0 will make it difficult to assess the role of the nonlinear mechanisms.

Second, the magnitude of the A_d (AH_0/L_e) parameter determines the significance of the nonlinear mechanisms for a particular test. Kabala et al. (1985) and McGuire and Zlotnik (1995) have shown that nonlinear mechanisms are of little significance when the magnitude of the A parameter is less than 1 and H_0 is less than 25% of L_e ($A_d < 0.25$). However, McElwee and Zenner (1993) obtained a value of 38 for the A parameter estimated from a series of slug tests in the same sand and gravel aquifer as in Figures 8.9 and 8.10. In that case, H_0 would have to be less than 0.7% of L_e for the conclusions of Kabala et al. (1985) and McGuire and Zlotnik (1995) to be appropriate, a condition that may often be difficult to meet in shallow groundwater investigations where L_e may not exceed a few meters. Although the details of the mechanisms that are responsible for the large magnitudes of the A parameter estimated from field data are still under investigation (McElwee and Butler, 1996), the significance of these mechanisms in high-conductivity formations is clear and test programs should be designed to assess their importance.

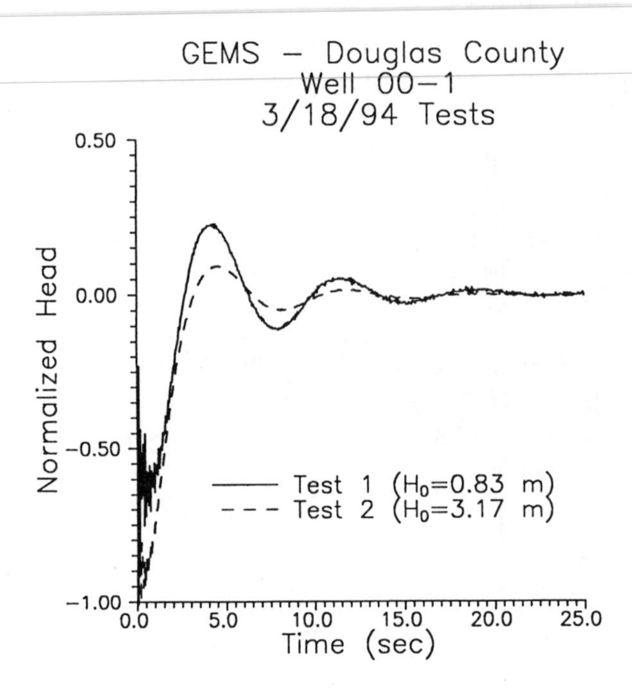

FIGURE 8.11 Normalized $(H(t)/(H_0)$ head vs. time plot of a pair of slug tests performed in well 00-1 at the Geohydrologic Experimental and Monitoring Site (GEMS) in Douglas County, Kansas (test 1 used in Figures 8.1, 8.7, and 8.8).

Third, even in cases where repeat slug tests indicate nonlinear mechanisms are of significance, the translation method can often be utilized to greatly lessen the impact of these mechanisms. For example, Figure 8.11 displays oscillating response data from two tests performed at the same well as used in the examples of the previous sections (Test 1 is utilized in those examples). Although the normalized plots clearly do not coincide, analysis of the two data sets with the linearized variant of the McElwee et al. method will yield hydraulic conductivity estimates that agree to within 20% if the time of the first peak in the data record is assumed to be the start time for each test. Since H_0 varied by a factor of over 3.8 between these tests, hydraulic conductivity estimates that agree to within 20% are an indication that the effect of nonlinear mechanisms on response data after the first peak is not great. This result, which has been verified by analyses of tests at other wells, indicates that the translation method both greatly lessens the impact of noninstantaneous slug introduction and removes much of the influence of nonlinear mechanisms from the data record. Thus, the translation method, in conjunction with one of the general linear models, is often a viable alternative to a more complicated nonlinear model.

Regardless of which model is used to analyze the response data, it may be virtually impossible to obtain a reasonable estimate of the hydraulic conductivity of the formation from slug tests performed in wells of very small diameter (i.e., considerably less than 0.05 m ID) screened in highly permeable material. The reason for this is that frictional losses within the well casing increase as the diameter of the casing decreases. This increase in frictional losses within the casing eventually results in a situation where the effective hydraulic conductivity of the casing is less than that of the formation. A slug test in such a well is basically just assessing the effective conductivity of the well itself, which may bear little resemblance to that of the formation. This effect can be particularly troublesome in multilevel slug tests performed with straddle packers, where long stretches of relatively small diameter tubing are often used between packers. McElwee and Butler (1996) provide field examples that demonstrate the impact of short stretches of 0.025 m ID and less tubing on response data from tests in a highly permeable aquifer. Thus, one must be skeptical of parameter estimates obtained from tests performed in formations of high hydraulic conductivity when the inner diameter of the test well (or tubing in the case of straddle-packer tests) is considerably less than 0.05 m and repeat test indicate that nonlinear mechanisms are significantly affecting the response data.

9 The Analysis of Slug Tests — Well Skins

CHAPTER OVERVIEW

All of the analysis methods discussed in the previous chapters have been based on the assumption that the formation is homogeneous, i.e., the hydraulic conductivity of the material immediately adjacent to the test interval is the same as the bulk average conductivity of the formation. However, as emphasized in Chapter 2, the process of drilling, installing, and developing a well will often result in the material in the immediate vicinity of the well having different characteristics than the formation as a whole. This zone of altered characteristics, the well skin, may have a considerable impact on the hydraulic conductivity estimate obtained from a slug test. A number of techniques have been developed for the analysis of response data from slug tests performed in the presence of well skins. The most common of these techniques are described in this chapter.

SLUG TESTS IN THE PRESENCE OF WELL SKINS

The possibility that the parameter estimates obtained from a slug test could be significantly affected by a well skin is not a new revelation. In one of the earliest discussions of the slug test, Ferris and Knowles (1963) cautioned that significant errors can be introduced into the conductivity estimate if the test well is inadequately developed, a warning that periodically has been repeated in the literature over the succeeding decades (e.g., Faust and Mercer, 1984; Butler et al., 1996). Several analysis methods have been developed to account for the effect of a well skin in response to these concerns. These techniques, which are primarily classified on the basis of the length of the well screen relative to the thickness of the formation (i.e., full vs. partial penetration), will be described in the following sections.

The hydraulic conductivity of a well skin can either be larger or smaller than that of the formation. The case of a skin of lower hydraulic conductivity, a low-K skin, is of most concern and, therefore, will be the primary emphasis of this discussion. Butler and co-workers (Hyder et al., 1994; Hyder and Butler, 1995; Butler, 1996) have shown that the impact of a high-K skin is quite small in most cases.

FULLY PENETRATING WELLS

The method of Ramey et al. (1975) is the most common approach for incorporating the effects of a well skin into the analysis of response data from slug tests performed in fully penetrating wells. The approximate deconvolution method of Peres et al. (1989) has also been used for this purpose (e.g., Chakrabarty and Enachescu, 1997).

Both of these methods will be described in the following sections. Karasaki (1990) has argued that the drillstem test presents several advantages for wells with low-conductivity skins; so, a brief discussion of the potential of that approach is also included.

The Ramey et al. Method

The Ramey et al. method is based on a mathematical model that can be defined as follows:

$$\frac{\partial^2 h}{\partial r^2} + \frac{1}{r}\frac{\partial h}{\partial r} = \frac{S_s}{K_r}\frac{\partial h}{\partial t} \quad (9.1a)$$

$$h(r,0) = 0,\ r_w < r < \infty \quad (9.1b)$$

$$H(0) = H_0 \quad (9.1c)$$

$$h(\infty,t) = 0,\ t > 0 \quad (9.1d)$$

$$h(r_w,t) - \sigma r_w \frac{\partial h(r_w,t)}{\partial r} = H(t),\ t > 0 \quad (9.1e)$$

$$2\pi r_w K_r B \frac{\partial h(r_w,t)}{\partial r} = \pi r_c^2 \frac{dH(t)}{dt},\ t > 0 \quad (9.1f)$$

where: σ = skin factor = $((K_r/K_{sk}) - 1)\ln(r_{sk}/r_w)$, [dimensionless];
K_{sk} = hydraulic conductivity of skin, [L/T];
r_{sk} = outer radius of skin, [L].

The effect of the well skin is incorporated into the model using Equation (9.1e), which is based on a steady-state representation of flow within the skin. The effects of the elastic storage properties of the skin are neglected, and the difference between the head change across the skin and the head change that would be seen over the same radial distance in the absence of a skin is assumed to be directly proportional to the flow across the skin, with σ, the skin factor, being the constant of proportionality. Since the transient effects produced by elastic storage mechanisms are neglected, the thickness of the skin is immaterial and the head change produced by the skin can be assumed to have occurred across a zone that is infinitely thin in its radial extent. This representation of flow within the well skin is therefore commonly known as the infinitely thin skin model.

As shown by Ramey and Agarwal (1972), the solution to the mathematical model defined in Equations (9.1a) to (9.1f) can be written as:

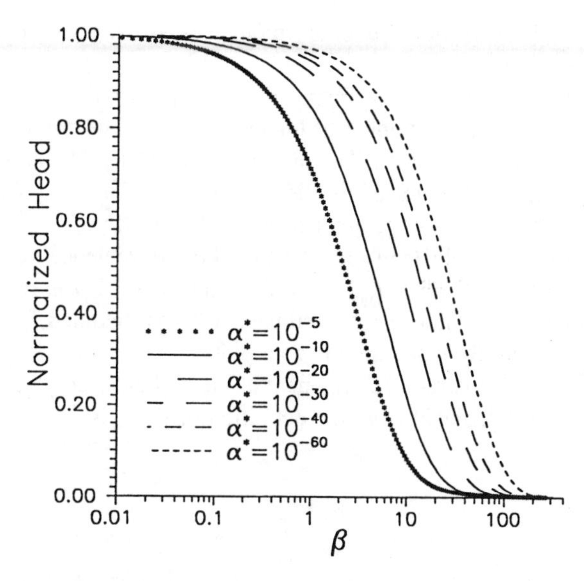

FIGURE 9.1 Normalized head ($H(t)/H_0$, where H(t) is deviation from static and H_0 is magnitude of the initial displacement) vs. the logarithm of β, the dimensionless time, type curves generated with the Ramey et al. model. Each curve labelled using α*, the modified dimensionless storage parameter (α* and β defined in text).

$$\frac{H(t)}{H_0} = f(\beta, \alpha, \sigma) \tag{9.2}$$

where α and β are as defined in Equation (5.2). Ramey et al. (1975) note that this three-parameter model is rather difficult to use in practice; so, they propose an approximate two-parameter model:

$$\frac{H(t)}{H_0} = f(\beta, \alpha^*) \tag{9.3}$$

where: $\alpha^* = \alpha e^{-2\sigma}$, [dimensionless].

This approximate solution, when plotted as normalized head vs. the logarithm of β, forms a series of type curves, with each type curve corresponding to a different value of α* (Figure 9.1). Note that α* is the same as the dimensionless storage parameter (α) of the previous chapters with the exception that the effective screen radius, r_w, has been replaced by $r_w e^{-\sigma}$. Thus, Equation (9.3) is simply the model of Cooper et al. (1967) in which the effective screen radius has been redefined using the skin factor.

The Ramey et al. method involves fitting one of the α* curves to the field data via manual curve matching or an automated analog in a manner similar to the method of Cooper et al. (e.g., Figure 5.3B). This technique consists of the following five steps:

1. The normalized response data are plotted vs. the logarithm of the time since the test began;
2. The data plot is overlain by a type-curve plot prepared on graph paper of the same format (i.e., number of log cycles). The type curves are moved parallel to the x axis of the data plot until one of the α^* curves approximately matches the plot of the field data. Note that the y axes are not shifted with respect to one another during this process;
3. Match points are selected from each plot. For convenience's sake, β is set to 1.0 and the real time ($t_{1.0}$) corresponding to $\beta = 1.0$ is read from the x axis of the data plot. An α^* estimate (α^*_{cal}) is obtained from the type curve most closely matching the data plot;
4. An estimate for the radial component of the hydraulic conductivity of the formation is calculated from the definition of β:

$$K_r = \frac{r_c^2}{Bt_{1.0}} \tag{9.4a}$$

5. An estimate for the specific storage of the formation is calculated from the definition of α^*:

$$S_s = \frac{\alpha^*_{cal}\, r_c^2}{r_w^2 e^{-2\sigma} B} \tag{9.4b}$$

where: α^*_{cal} = α^* value calculated via curve matching.

There are several issues of practical importance with respect to the method of Ramey et al. First, Equation (9.3) is an approximation of the more involved solution given in Equation (9.2), which is itself based on an approximate representation of flow within the well skin (i.e., the infinitely thin skin model). Moench and Hsieh (1985a,b) have shown that the approximations underlying Equations (9.2) and (9.3) are reasonable for conventional slug tests. However, in shut-in tests, the thickness of the well skin and its elastic storage properties cannot be ignored; so, a more rigorous finite-thickness skin model must be considered (Moench and Hsieh, 1985a).

Ramey et al. (1975) propose three different curve-matching schemes for the estimation of formation parameters. The most commonly used approach is the one described in the preceding paragraphs, which employs the same semilog plot as the Cooper et al. method. This semilog plot, which is dubbed the "Ramey type a" plot by petroleum engineers, emphasizes response data in the range of normalized heads commonly measured in groundwater applications. The other two curve-matching schemes are based on log-log plots that emphasize very late-time (Ramey type b plot) or very early-time (Ramey type c plot) response data, and thus are of rather limited use in most shallow groundwater investigations.

One of the primary issues of practical concern with respect to the method of Ramey et al. is that of the uncertainty in the hydraulic conductivity estimate. This uncertainty is primarily produced by the nonuniqueness of the α^* type curve match. Close scrutiny of Figure 9.1 will reveal that the shape of the type curves becomes very similar as α^* decreases, making it extremely difficult, on the basis of shape alone, to distinguish between α^* curves that differ by five to ten orders of magnitude. When sensor and other background noise is superimposed on these plots, there is little chance of obtaining a reasonable α^* estimate. Ostrowski and Kloska (1989) propose that head derivative plots be used to obtain a better α^* estimate. Although derivative plots accentuate the subtle differences between the type curves, background noise and uncertainty about H_0 make it difficult to exploit the theoretical advantages of this approach. However, despite the nonuniqueness of the α^* estimate, Moench and Hsieh (1985b) show that a conductivity estimate within an order of magnitude of the actual hydraulic conductivity of the formation should be obtainable from most slug tests performed in fully penetrating wells with low-K skins.

A closely related issue is that of the quality of the specific storage estimate. Unfortunately, it will be virtually impossible to obtain a reasonable estimate of the specific storage of the formation from a test in a well with a low-K skin. There are two reasons for this situation. First, the α^* estimate will undoubtedly be in error by several orders of magnitude as a result of the nonuniqueness discussed in the previous paragraph. Second, even if a reasonable estimate of α^* is obtained, uncertainty about the magnitude of the skin factor will make it very difficult to calculate an accurate specific storage estimate from Equation (9.4b). In most cases, estimates of specific storage based on lithologic information and the judgment of experienced field hydrologists will be superior to what can be produced from an analysis with the Ramey et al. method.

An additional issue of practical importance is the ramifications of the manner in which the well skin is represented in the Ramey et al. model. In this model, the well skin is assumed to extend along the entire screened interval in a manner similar to that depicted in Figure 2.4A. In reality, however, such conditions may not be commonplace. If the well has been developed to any degree, one would expect conditions more similar to those depicted in Figure 2.4B. If layering has imbued the formation with a considerable degree of vertical anisotropy, distinguishing conditions depicted in Figure 2.4B from the case of a fully penetrating well in a homogeneous formation may be quite difficult. In that situation, the conductivity estimate obtained from the analysis will underpredict the formation conductivity by a factor that is a ratio of the nominal screen length over the thickness of the developed zones. If the formation is essentially isotropic with respect to hydraulic conductivity, the existence of a few developed channels through the well skin will produce a significant component of vertical flow. However, the head loss produced by this vertical flow can be represented as an additional skin effect (the "pseudoskin" of petroleum engineering); so, the test can still be analyzed using the Ramey et al. method. The hydraulic conductivity estimate obtained from this analysis will be of similar quality to that obtained for the case of a well skin extending along the entire screened interval.

A related issue is that of the ramifications of a dynamic skin. The Ramey et al. method is, as are all analysis techniques that incorporate the effects of a well skin, based on the assumption that the skin is static in nature. If the skin behaves in a dynamic fashion, such as shown in Figure 2.2, it may be very difficult to obtain a reasonable estimate of the hydraulic conductivity of the formation from an analysis of the response data. Thus, as emphasized in Chapter 2, it is imperative that dynamic-skin effects be recognized through a program of repeat slug tests.

An additional issue of practical importance is that of the analysis of shut-in slug tests. Moench and Hsieh (1985a) have demonstrated that the infinitely thin skin model of Ramey et al. is not appropriate for shut-in tests. Instead, they propose a finite-thickness skin model as the most appropriate approach for the analysis of shut-in test data. They caution, however, that it can be extremely difficult to obtain a reasonable estimate of the hydraulic conductivity of the formation using this model. Thus, hydraulic conductivity estimates obtained from shut-in tests must be viewed with considerable skepticism if there is a possibility that a low-K skin could have affected the response data.

A final issue of practical importance is that of the relationship between the Ramey et al. method and analysis techniques developed for homogeneous formations. A major advantage of the Ramey et al. method is that it is a simple extension of the technique of Cooper et al. The same curve matching approaches and type curves are used in both methods. The only difference between the two methods is in the equations for estimation of specific storage (Equations 5.4 and 9.4b). Further insight can be obtained by rewriting Equation (9.4b) in the same form as the right-hand side of Equation (5.4):

$$S_s e^{-2\sigma} = \frac{\alpha^*_{cal}\, r_c^2}{r_w^2 B} \tag{9.5}$$

where: $S_s e^{-2\sigma}$ = apparent specific storage, [1/L].

The form of Equation (9.5) demonstrates that if the Cooper et al. method is applied to test data from a well with a skin, an apparent specific storage estimate will be obtained whose magnitude is a function of the skin factor term. Thus, the existence of a well skin should be evident from the physically implausible value obtained for the specific storage estimate. In a fully penetrating well, a physically implausible specific storage estimate is one of the best indications of the existence of a well skin. It is important to emphasize that the presence of a well skin will be very difficult to recognize using the Hvorslev method. As discussed in Chapter 5, the Hvorslev method is based on a theoretical development in which it is assumed that a plot of log normalized head vs. time will be linear in the homogeneous case. Actually, however, this plot will be most nearly linear in the case of a low-conductivity skin; so, there may be considerable confusion about the significance of a linear plot for slug tests performed in fully penetrating wells. Unfortunately, even if the presence of a low-K skin is strongly suspected, it may be impossible to obtain a reasonable estimate of the hydraulic conductivity of the formation using the Hvorslev method (Hyder et al., 1994).

The Peres et al. Approximate Deconvolution Method

Although the existence of a low-conductivity skin can often be recognized from the results of an analysis using the Ramey et al. method, the conductivity estimate will only be a rather crude approximation of the actual hydraulic conductivity of the formation. The deconvolution method of Peres et al. (1989) presents an alternate method for the analysis of data from tests in wells with low-K skins. In fully penetrating wells, this method can often produce a reasonable estimate of the hydraulic conductivity of the formation.

As described in Chapter 5, the deconvolution method is based on transforming response data from a slug test into the equivalent drawdown that would have been produced by a constant rate of pumping at the same well in the absence of wellbore storage effects. Mishra (1991), in one of the earliest presentations on the approach, employs a hypothetical example to demonstrate that the deconvolution method can be used to transform response data from a slug test performed in the presence of a finite-thickness skin into the equivalent drawdown that would be produced by pumping from a well at the center of a small disk embedded in a matrix of higher conductivity. As has been shown by Butler (1988), among others, an excellent estimate of the hydraulic conductivity of the formation can be obtained from a pumping test in this configuration by applying the Cooper-Jacob method (Cooper and Jacob, 1946; Kruseman and de Ridder, 1990) to the late-time drawdown data. Mishra (1991) demonstrates that the Cooper-Jacob approach can also be applied to the transformed response data to obtain a quite reasonable estimate of the conductivity of the formation.

A series of hypothetical examples can be used to demonstrate the potential of the deconvolution approach. Figure 9.2A is a plot of responses from two hypothetical slug tests performed in a well with a low-conductivity skin ($K_{sk}/K_r = 0.05$, S_s is constant) simulated using the finite-thickness skin model of Moench and Hsieh (1985a). Simulated responses from a hypothetical test in a homogeneous formation are also included for comparison. Figure 9.2B is the equivalent drawdown produced by transforming the simulated responses using Equation (5.13b). The near-parallel drawdown plots at moderate to large times are an indication that changes in the equivalent drawdown at these times are primarily a function of the conductivity of the formation, i.e., properties of the skin have essentially no influence on changes in drawdown over that time interval. Application of the Cooper-Jacob method to the simulated response plots of Figure 9.2B over a normalized head range usually available in shallow groundwater investigations (0.01 to 0.30) produced conductivity estimates that ranged between 0.60 and 0.82 of the actual formation value. However, if the assumption is made that noise-free data are available to a normalized head of 0.0001, a conductivity estimate within 5% of the actual formation value can be obtained. This result highlights the potential of the approximate deconvolution method for petroleum engineering applications, where the H_0 tend to be very large.

There are three issues of practical importance with respect to the approximate deconvolution method. First, as with the case of tests in homogeneous formations, the deconvolution method requires that recovery data be closely spaced in time. Data

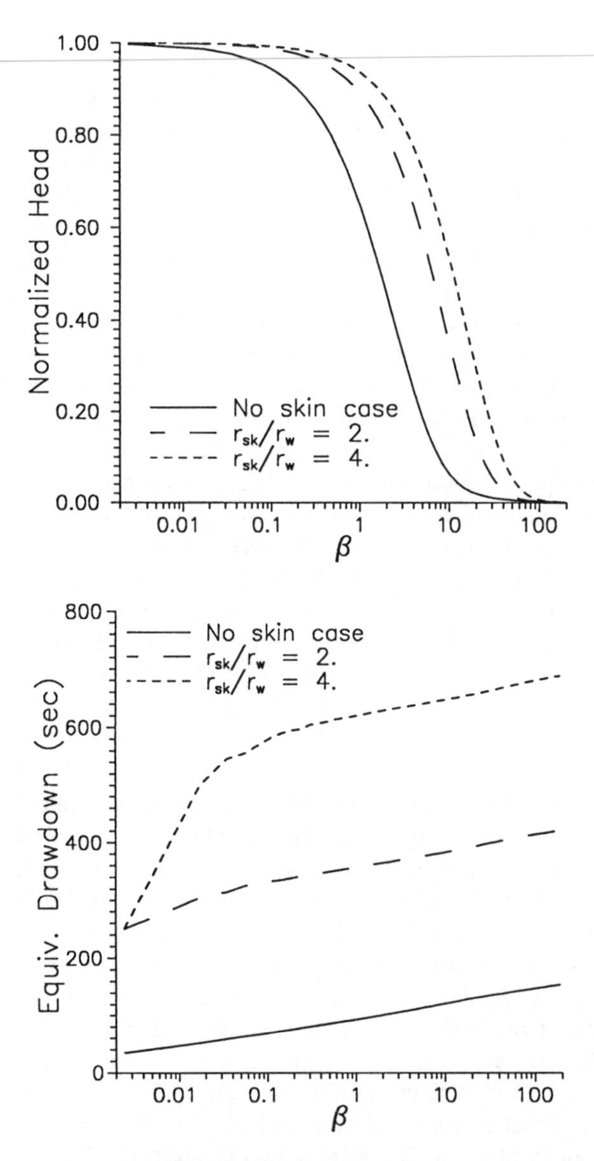

FIGURE 9.2 (A) Normalized head ($H(t)/H_0$) vs. log β plot of a series of hypothetical slug tests simulated with the solution of Moench and Hsieh; (B) Equivalent drawdown (after removal of wellbore storage effects) plot for simulated responses of Figure 9.2A (r_{sk} and the equivalent drawdown term are defined in text).

sets in which the normalized head changes by more than a few percent between adjacent measurements can be difficult to analyze. Similarly, better estimates of the conductivity of the formation can be obtained if a large H_0 is used, high accuracy pressure transducers are available, background noise is quite small, and the test is run to complete recovery.

The results displayed in Figure 9.2B indicate that the deconvolution approach has considerable potential for the analysis of response data from slug tests performed in fully penetrating wells. This is particularly true when the well skin extends along the entire screened interval. The technique may be less successful when conditions are similar to those depicted in Figure 2.4B. In a formation with a very significant layering-induced anisotropy, the limitations of the approach will be similar to those encountered with the Ramey et al. method. In the case of a near-isotropic formation, the vertical flow produced by the channels through the well skin may change the time of onset of the semilog straight line. However, a reasonable estimate of the conductivity of the formation should still be possible.

Finally, the existence of a low-conductivity skin may often be evident from the two-limb, concave-downward character of the plot (e.g., upper curve in Figure 9.2B). Since the slope of the plot will be decreasing as it converges on the slope that would be obtained in the homogeneous case, the conductivity estimate will always be lower than the conductivity of the formation. However, that estimate should be within a factor of two to three of the formation value in most cases.

The Drillstem Test

Karasaki (1990) demonstrates that a drillstem test can potentially be quite useful in wells with low-conductivity skins. He shows that a very reasonable estimate of the hydraulic conductivity of the formation can be obtained with the Theis recovery analysis. Although not emphasized by Karasaki (1990; 1991), it is important to note that the normalized head at which the dst responses will converge on the theoretical straight line may be much smaller in the presence of a low-conductivity skin than that shown in Figures 7.8 to 7.10. For example, in the plots of field data given in Karasaki (1991), convergence does not occur until the normalized head is considerably less than 0.02. Thus, similar to the approximate deconvolution method, the viability of this approach will depend on the magnitude of the contrast between the conductivity of the formation and that of the skin. The greater the magnitude of the contrast, the less likely the convergence will occur over the normalized head ranges commonly used in shallow groundwater investigations.

Partially Penetrating Wells

The effects of a well skin are incorporated into the analysis of a slug test performed in a partially penetrating well using one of three approaches: the method of Ramey et al. (1975), the KGS model (Hyder et al., 1994) and associated approaches (i.e., the model of Dougherty and Babu, 1984), and the approximate deconvolution method of Peres et al. (1989). Each of these techniques are described in the following sections.

The Ramey et al. Method

The method of Ramey et al. is the most commonly used approach for incorporating the effects of a well skin into the analysis of response data from slug tests performed in partially penetrating wells. The Ramey et al. method for the case of a partially penetrating well is again based on the mathematical model defined in Equations (9.1a) to (9.1f). Thus, although the well is only screened over a portion of the formation, the slug-induced flow is considered to be purely radial in nature. The head loss produced by the vertical component of flow is represented as an additional skin effect (the "pseudoskin" of petroleum engineering) and lumped together with the head losses produced by the actual well skin. The steps employed for the analysis are exactly the same as those used in the case of a fully penetrating well.

The same issues of practical importance with respect to the Ramey et al. method for fully penetrating wells are also of significance for tests performed in partially penetrating wells. An additional issue of particular significance for partially penetrating wells is that of the effective screen length. Since the Ramey et al. method is based on a mathematical model that assumes all flow is purely radial and lumps vertical-flow induced head losses with skin-induced losses, the effective screen length is not the nominal screen length of the well but is the thickness of the unit being tested. Use of the nominal screen length for this parameter will result in an underestimation of hydraulic conductivity by a factor that is the ratio of the actual unit thickness over the nominal screen length.

A closely related issue is that of the estimation of the actual thickness of the unit or interval through which the slug-induced flow occurs (the flow interval). Uncertainty about this thickness is greatest in wells that are screened for a relatively short length in the upper portions of what is suspected to be a much thicker unit. Butler and Healey (1998) discuss a field example that demonstrates the error that can be introduced into the conductivity estimate through uncertainty about the actual thickness of the flow interval. Information from drilling logs and geophysical surveys is often invaluable in providing constraints on this parameter.

A final issue of practical importance is that of the significance of the estimate of the $S_se^{-2\sigma}$ term obtained using the Ramey et al. method. In the case of a fully penetrating well, a value much below 10^{-6} m^{-1} is an indication of a low-conductivity well skin. However, in a partially penetrating well, a very low value can also be obtained for this quantity as a result of head losses produced by vertical flow. Thus, relatively little significance can be attributed to a low $S_se^{-2\sigma}$ value obtained from a slug test in a partially penetrating well analyzed with the method of Ramey et al.

The KGS Model

For slug tests performed in partially penetrating wells, the best indication of the existence of a well skin can be obtained from the results of an analysis using the KGS model (Hyder et al., 1994) or its forerunner, the model of Dougherty and Babu (1984). The KGS model is the preferred approach because of its greater flexibility, i.e., the inclusion of vertical anisotropy and both confined and unconfined conditions.

The KGS model for confined formations is based on the mathematical model defined in Equations (5.24a) to (5.24g), while the unconfined variant is based on a similar model defined in Equations (6.7a) to (6.7h). Finite-thickness skin variants of the model are available (Hyder et al., 1994) and the model can also be readily modified to incorporate the infinitely thin skin conceptualization employed by Dougherty and Babu (1984). However, variants of the model that include skin effects are really not necessary for practical applications, as the homogeneous form of the model will be adequate to indicate the presence of a well skin.

The steps employed for the analysis of the response data are exactly the same as those described in Chapters 5 and 6 for tests in homogeneous formations. The emphasis of the analysis, however, is now on the magnitude of the apparent specific storage estimate. If the specific storage estimate is larger than expected for an isotropic formation, there is a strong possibility that the response data are being affected by a high-conductivity well skin. Another possibility is an anisotropy ratio (K_z/K_r) that is smaller than expected. Fortunately, a preliminary analysis with the Cooper et al. method, such as described in Chapter 5, can often help clarify which mechanism is of most significance. If a good match between a theoretical type curve and the field data is only possible with an extremely low value for specific storage, there is a very strong likelihood that the response data are being affected by a low-conductivity skin. As in the case of the Cooper et al. method, a physically implausible S_s estimate is the result of the homogeneous variant of the KGS model attempting to account for the additional head loss produced by a low-conductivity skin. Unfortunately, regardless of which variant of the KGS model is employed, the hydraulic conductivity estimate obtained from the analysis of response data affected by a low-conductivity skin will only be a very rough approximation of the formation conductivity as a result of the nonuniqueness of the type curve match.

An implausibly low S_s estimate obtained with the KGS model is one of the best indications of a low-conductivity well skin for tests performed in partially penetrating wells. However, the KGS model is not the most commonly used technique for the analysis of response data from tests performed in partially penetrating wells. The most commonly used approaches are the Hvorslev and Bouwer and Rice methods, both of which are based on a quasi-steady-state conceptualization of slug-induced flow. According to this conceptualization, a plot of the logarithm of normalized head vs. time will be linear in a homogeneous formation. Actually, however, the linearity of this plot does not depend on the assumption of homogeneity. When data from tests performed in partially penetrating wells of moderate to small aspect ratios are used, this plot will be quite linear both in the case of a homogeneous formation and that of a low-conductivity well skin. The result is that a low-conductivity skin will be extremely difficult to recognize with these methods. Unfortunately, theoretical investigations have shown that it may be virtually impossible to obtain a reasonable estimate of the hydraulic conductivity of the formation when quasi-steady-state techniques are used to analyze response data from wells with low-conductivity skins (Hyder et al., 1994; Hyder and Butler, 1995).

The Peres et al. Approximate Deconvolution Method

The techniques described in the previous two sections will, at best, provide some indication of the presence of a low-conductivity skin and a rough estimate of the conductivity of the formation. As was the case for tests in fully penetrating wells, the approximate deconvolution method of Peres et al. (1989) may be the best option for obtaining a reasonable estimate of the hydraulic conductivity of the formation from tests performed in wells with low-conductivity skins. The Peres et al. method is applied to data from partially penetrating wells in the same manner as for data from fully penetrating wells. The resulting conductivity estimate, however, is not the average conductivity of the screened interval, but instead is the average conductivity of the flow interval.

The major issues of practical importance are similar to those discussed in earlier sections on the approximate deconvolution method. In this case, however, less significance can be attributed to the form of a concave-downward equivalent drawdown plot. For tests in fully penetrating wells, a concave-downward plot (e.g., Figure 9.2B) is an indication of the existence of a low-conductivity skin. However, as explained in Chapter 5, the vertical component of flow produced by partial penetration will also imbue the equivalent drawdown plot with a concave-downward curvature. Thus, some care should be given to the interpretation of the concave-downward curvature. Regardless of the mechanism, the concave-downward curvature will always lead to an underestimation of hydraulic conductivity.

10 The Analysis of Slug Tests — Multiwell Tests

CHAPTER OVERVIEW

All of the analysis techniques presented in the previous chapters have used data from the test well to obtain estimates of the transmissive and storage properties of the formation. If observation wells are available, however, response data from those wells can be the source of valuable additional information about the hydraulic properties of the formation. In this chapter, the major techniques for the analysis of response data from observation wells are described.

MULTIWELL SLUG TESTS

Traditionally, slug tests have been performed using a single well (designated here the test well) as both the site of the stress and the site at which head measurements are taken. This practice is in keeping with the common perception that a slug test only affects a very small volume of the formation (e.g., Ferris and Knowles, 1963; Rovey and Cherkauer, 1995). Field data from slug tests with observation wells, however, indicate that a slug test may actually be affecting a much larger volume of the formation than is commonly thought. Figure 10.1 is a plot of normalized head vs. the logarithm of time for a slug test performed at a monitoring site in Lincoln County, Kansas. In this case, the head was measured at an observation well 6.45 m from the test well. Figure 10.2 is a similar plot for a test performed at a monitoring site in Stanton County, Kansas, where the observation well was 10.2 m from the test well. Although in both of these cases the interval being tested was a confined semiconsolidated sandstone, such behavior is certainly not restricted to that geology. Novakowski (1989), for example, presents results from several tests performed in a fractured shale using an observation well 15 m from the test well, while Spane et al. (1996) discuss a series of tests in unconsolidated sands and gravels in which the distance to the observation well ranged between 9 and 11 m. The distances involved in these examples demonstrate that the parameter estimates obtained from a slug test may often be reflective of bulk formation properties, and not just material in the immediate vicinity of the test well.

Despite the common perception that the volume of the formation affected by a slug test is quite small, the field data discussed in the preceding paragraph are consistent with theoretical results that have been reported in the literature for over two decades. Ramey et al. (1975), using an extension of the Cooper et al. (1967) solution, were the first to demonstrate that slug-induced head disturbances can propagate far into the formation. The results of their analysis indicate that responses should be measurable at distances over several hundred times the effective screen

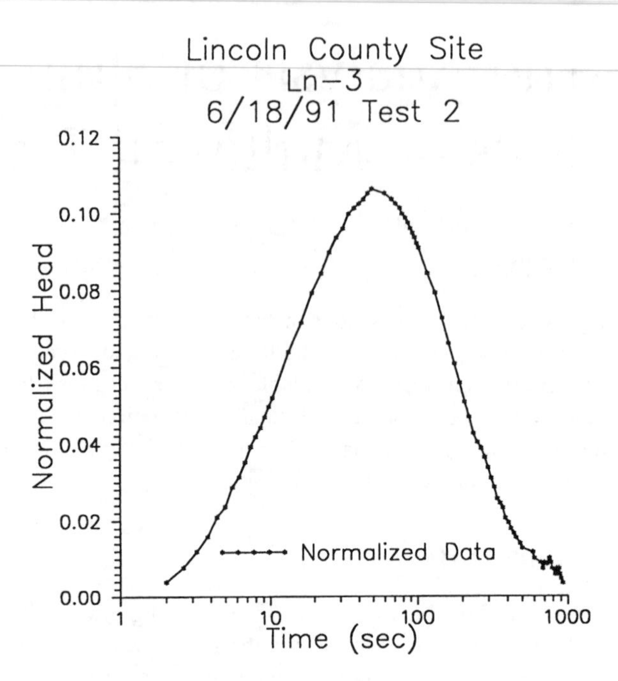

FIGURE 10.1 Normalized head ($h_{ow}(t)/H_0$, where $h_{ow}(t)$ is deviation from static at an observation well and H_0 is magnitude of the initial displacement) vs. logarithm of time plot for a multiwell slug test performed at a monitoring site in Lincoln County, Kansas (test initiated in well Ln-2, h_{ow} measured in well Ln-3 at a distance of 6.45 m from Ln-2).

radius (r_w) of the test well. The initial theoretical work of Ramey et al. was later expanded upon by Barker and Black (1983), Sageev (1986), and Karasaki et al. (1988), among others. This body of work demonstrated that the size of the volume of the formation affected by a slug test is most dependent on the dimensionless storage parameter (α). Figure 10.3 is a dimensionless semilog plot of theoretical responses at an observation well a distance of 100 r_w from the test well simulated using the solution of Cooper et al. (1967). As shown in the figure, the smaller the α, the larger the normalized response and, therefore, the greater the volume affected by the test.

The actual volume affected by a slug test is difficult to define. Most workers have used a definition based on the distance some arbitrary normalized head (usually between 0.01 and 0.10) will propagate during a test. Barker and Black (1983), for example, propose an approximate equation for the distance a normalized head of 0.10 will propagate during a slug test initiated in a fully penetrating well, which can be written for a porous formation as:

$$R_{D\max} = \alpha^{-0.5} \tag{10.1}$$

where: $R_{D\max}$ = the maximum radial extent of a slug test divided by the effective screen radius of the test well, [dimensionless].

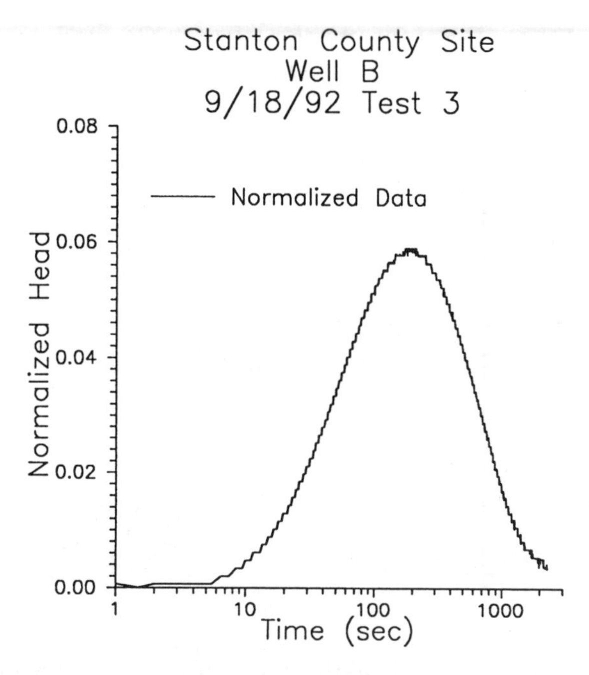

FIGURE 10.2 Normalized head ($h_{ow}(t)/H_0$) vs. logarithm of time plot for a multiwell slug test performed at a monitoring site in Stanton County, Kansas (test initiated in well A, h_{ow} measured in well B at a distance of 10.2 m from well A).

The approximate relationship expressed in Equation (10.1) is depicted graphically in Figure 10.4. Barker and Black (1983) point out, however, that the parameter estimates obtained from slug tests are spatially weighted averages of formation properties. Thus, the further the distance from the test well, the smaller the contribution of that portion of the formation to the parameter estimate. Eventually, there will be a point at which the contributions of the portions of the formation lying radially outward from that point can be ignored. Although the numerical investigation reported by Harvey (1992) indicates that Equation (10.1) is a reasonable representation of the distance to that point, other authors have proposed alternate relationships (e.g., Guyonnet et al., 1993). However, regardless of the specifics of the various relationships that have been proposed, they all indicate that the size of the volume affected by a slug test is large enough that a slug test may be a reasonable alternative to a pumping test when the major purpose of the test is to obtain an estimate of bulk formation properties, and when the well-formation configuration has a small to moderate value of α. Obviously, a slug test cannot be considered as a viable alternative to a pumping test if information about the hydraulic boundaries of a flow system is desired.

The use of observation wells with slug tests (henceforth designated multiwell slug tests) would be of relatively little interest if the sole purpose of the wells was to demonstrate the size of the volume affected by a test. Karasaki et al. (1988) were the first to recognize that the real advantage of a multiwell slug test is that it can

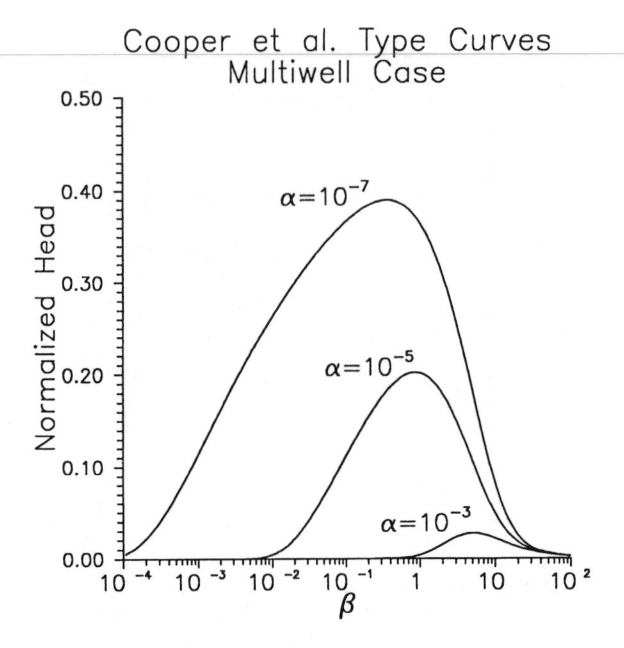

FIGURE 10.3 Normalized head ($h_{ow}(t)/H_0$) vs. logarithm of β type curves generated with the Cooper et al. model (observation well at a distance of $100r_w$ from the test well ($r_D = 100$)).

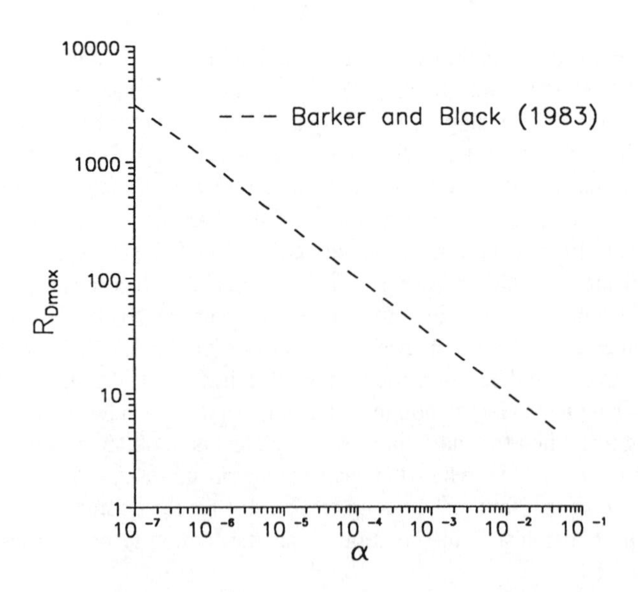

FIGURE 10.4 Logarithm of the dimensionless maximum radial extent of a slug test vs. logarithm of α (R_{Dmax} defined in text).

provide additional information beyond that obtainable from a conventional single-well test. In the case of a test in a fully penetrating well, the use of response data from an observation well allows a very reasonable estimate of the dimensionless storage parameter (α) to be obtained. Figure 10.3 displays a series of type curves generated using the solution of Cooper et al. (1967) for an observation well located at a distance of 100 r_w from the test well. In the case of a single-well test (e.g., Figure 5.2), the similarity in the shape of the type curves introduces a great deal of uncertainty into the α estimate. However, as shown in Figure 10.3, the type curves for the multiwell case are sufficiently different in terms of both shape and magnitude that a very reasonable estimate of α can be obtained. Karasaki et al. (1988) and McElwee et al. (1995b) provide further discussion of this issue. Note that multiwell slug tests have also been referred to as "pulse tests" (e.g., Novakowski, 1989) and "slug interference tests" (e.g., Chu and Grader, 1991; Spane et al., 1996) in the groundwater and petroleum-engineering literature. The pulse-test designation is based on the similarity of the responses at the observation well to those produced by cyclical variations in pumping rate (e.g., Johnson et al., 1966; Walter and Thompson, 1982). The term slug-interference test is in keeping with the use of "interference" in the petroleum-engineering literature to indicate a hydraulic test with observation wells, i.e., a test in which a head disturbance at one well "interferes" with the head response at a nearby well. Since neither of these alternatives clearly convey the idea that head data are collected at locations other than the test well, the term multiwell slug test is used here.

The discussion of the preceding paragraphs pertained to tests performed in fully penetrating wells. However, most slug tests for environmental applications are performed in partially penetrating wells. In this case, slug-induced responses are also a function of the ψ ($(K_z/K_r)^{1/2}/(b/r_w)$) parameter. As discussed in Chapter 5, it is virtually impossible to obtain reasonable estimates of specific storage and the anisotropy ratio from single-well slug tests performed in partially penetrating wells. Use of an observation well, however, can greatly improve the situation. Spane et al. (1996) present the results of a series of field tests in partially penetrating wells in an unconfined aquifer, and examine the sensitivity of responses at observation wells to variations in specific storage and the anisotropy ratio. Butler and McElwee (1996) use results from a series of field tests in a confined aquifer to demonstrate the impact of anisotropy on responses at an observation well. The results of these analyses indicate that considerable information can be obtained from multiwell slug tests initiated in partially penetrating wells.

Although several methods have been proposed for the analysis of response data from multiwell slug tests, none have been extensively applied in the field. The focus of this discussion will be on two methods developed for tests in fully penetrating wells: the multiwell extension of the method of Cooper et al. (1967) and the method of Chu and Grader (1991). These two techniques are described in the following sections. After the description of these techniques, possible approaches for tests in partially penetrating wells are briefly examined.

The Cooper et al. Method

The Cooper et al. method is based on the mathematical model defined in Equations (5.1a) to (5.1f). In the case of a multiwell test, the analytical solution for head at an observation well ($h(r_D,t)$) can be written as:

$$\frac{h(r_D,t)}{H_0} = f(\beta,\alpha) \quad (10.2)$$

where: $r_D = r_{obs}/r_w$, [dimensionless];
r_{obs} = radial distance from test well to observation well, [L].

This solution, when plotted as normalized head vs. the logarithm of β, forms a series of type curves, with each type curve corresponding to a different value of α (e.g., Figure 10.3). As in the single-well case, the Cooper et al. method involves fitting one of the α curves to the field data via manual curve matching or an automated analog. The method essentially consists of the following five steps:

1. The normalized response data are plotted vs. the logarithm of the time since the test began;
2. The data plot is overlain by a type-curve plot prepared on graph paper of the same format (i.e., number of log cycles) for an observation well at that dimensionless distance from the test well. The type curves are then moved parallel to the x axis of the data plot until one of the α curves approximately matches the plot of the field data. Note that the y axes are not shifted with respect to one another during this process;
3. Match points are selected from each plot. For convenience's sake, β is set to 1.0 and the real time ($t_{1.0}$) corresponding to $\beta = 1.0$ is read from the x axis of the data plot. An α estimate (α_{cal}) is obtained from the type curve most closely matching the data plot;
4. An estimate for the radial component of hydraulic conductivity is calculated from the definition of β:

$$K_r = \frac{r_c^2}{Bt_{1.0}} \quad (10.3a)$$

5. An estimate for the specific storage is calculated from the definition of α:

$$S_s = \frac{\alpha_{cal} r_c^2}{r_w^2 B} \quad (10.3b)$$

where: α_{cal} = α value calculated via curve matching.

There are several issues of practical importance for the multiwell variant of the Cooper et al. method. First, the mathematical model upon which the method is based assumes that the head is being measured in the formation. In reality, the head in an observation well may be somewhat different from that which would be measured in the formation in the absence of the well as a result of the finite diameter of the observation well (i.e., wellbore storage) and skin effects. Novakowski (1989) demonstrates the magnitude of the error that can be introduced into parameter estimates as a result of the failure to consider wellbore storage, and recommends that wellbore storage effects be minimized by use of a simple packer and transducer arrangement in the observation well (e.g., Figure 10.5). If such an arrangement is not used, Novakowski (1989) proposes an approximate graphical approach for incorporating the effect of wellbore storage in the observation well into the analysis. A more rigorous consideration of wellbore storage at the observation well can be obtained using the method of Chu and Grader (1991) described in the following section. However, the difference between these two approaches is quite small for an observation well at a moderate to large distance from the test well. Despite the availability of solutions that incorporate wellbore storage at the observation well, a packer-transducer arrangement should be used in the observation well if at all possible. As will be discussed further in the following section, a very significant advantage of a packer in the observation well is that the impact of a low-conductivity skin at that well is greatly diminished. Thus, if an observation well is packed off, conditions at the observation well should have a very limited impact on parameter estimates obtained using the method of Cooper et al.

A second issue of considerable practical importance is the impact of a well skin at the test well. Figure 10.6 demonstrates how a skin at the test well can affect responses at an observation well. Although a high-conductivity skin has a very limited impact on the response data, a low-conductivity skin will lag and damp responses, resulting in an underestimation of hydraulic conductivity and an overestimation of the dimensionless storage parameter. Thus, it is extremely important that the existence of a low-conductivity skin at the test well be detected in the course of the analysis. An effective approach for doing this is to simultaneously analyze data from both the test and observation wells. For example, type curves from Figures 5.2 and 10.3 can be combined into a single set of type curves and the response data from both wells can be placed on a single plot (e.g., Figure 10.7). The type curves can then be moved along the x axis of the combined data plot until the response data are matched by a single set of α curves. If a low-conductivity skin is present at the test well, a combined match will not be possible. The data from the test well will require an α estimate that is much lower than that of the formation, while the data from the observation well will require an α estimate that is higher than that of the formation. The effect of a high-conductivity skin will be much less dramatic. In that case, data from the test well will require a higher than expected α estimate, while the α estimate obtained from an analysis of data from the observation well will essentially be unaffected by the skin. Thus, simultaneous analysis of data from the test well and all observation wells is strongly recommended. If the analysis

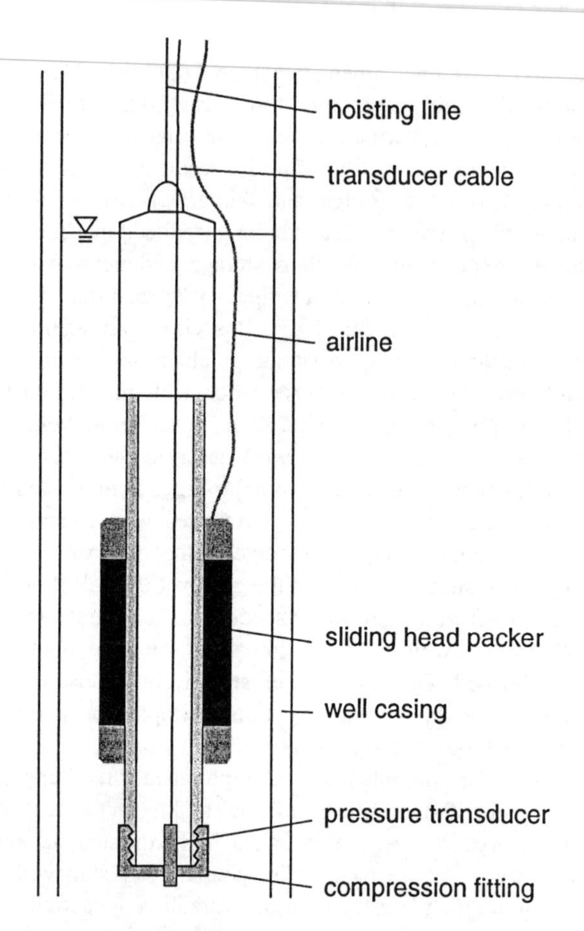

FIGURE 10.5 Schematic of a packer and transducer arrangement for use in an observation well (transducer measures pressure below packer; apparatus shown with uninflated packer; figure not to scale).

indicates that a low-conductivity skin may be present at the test well, the method of Chu and Grader should be selected for the analysis of the response data.

A field example can demonstrate the use of the multiwell variant of the Cooper et al. method. In May and June of 1991, a series of multiwell slug tests were performed at a monitoring site in Lincoln County, Kansas (Butler and Liu, 1997). The test well (Ln-2) and the observation well (Ln-3) were both screened in the semiconsolidated upper deltaic sand unit of the Dakota Formation. Table 10.1 summarizes the well-construction information for the two wells. Table 10.2 lists the response data employed in the analysis, while Figure 10.7 is a plot of that data in the normalized head vs. log time format of the Cooper et al. method. A packer and transducer arrangement similar to that in Figure 10.5 was used in the observation well to minimize the impact of wellbore storage and skin effects.

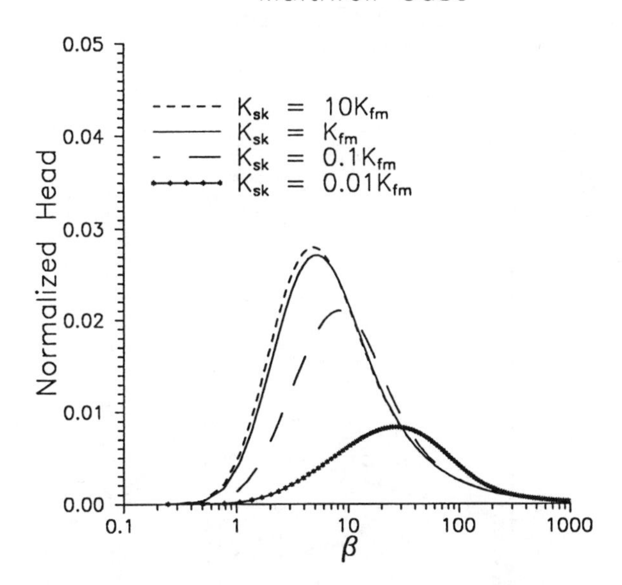

FIGURE 10.6 Normalized head ($h_{ow}(t)/H_0$) vs. logarithm of β plots generated with the Moench and Hsieh finite-thickness skin model (observation well at a distance of 100 r_w from the test well ($r_D = 100$); $r_{sk} = 2r_w$).

Figure 10.8 displays the results of the curve-matching process used in the Cooper et al. method. A very close match was found between the data from both wells and the $\alpha = 2.3 \times 10^{-4}$ type curves. The excellent agreement between type curves for a single α and the response data from both wells is a convincing demonstration of the absence of a low-conductivity skin at the test well. Using the match points from the simultaneous fit in Equations (10.3a) and (10.3b) yields K_r and S_s estimates of 1.06 m/d and 8.5×10^{-6} m^{-1}, respectively, both of which are quite reasonable values for a semiconsolidated sand unit. Note that in this analysis the effective radius of the well screen was set equal to the outer radius of the filter pack (0.102 m). The impact of a high-conductivity skin can be demonstrated by using the nominal radius of the well screen (0.051 m) for the effective radius. In this case, the filter pack acts as a high-conductivity skin and there is a difference of a factor of four between the α value from the type curve fit to data from the test well and that from the fit to data from the observation well.

The response data from the series of multiwell tests performed at this site can also be used to demonstrate the impact of wellbore storage and a low-conductivity skin at the observation well. Figure 10.9 displays data from a pair of tests performed with (6/18/91) and without (5/21/91) a packer in the observation well. These two tests clearly show how wellbore storage at the observation well will lag and damp responses. Note that the response data from the test well approximately coincide for

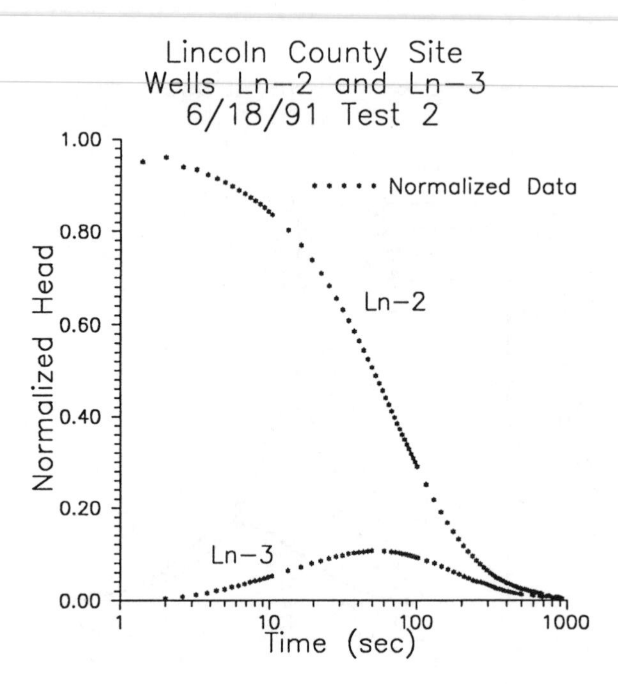

FIGURE 10.7 Normalized head ($H(t)/H_0$ and $h_{ow}(t)/H_0$) vs. logarithm of time plot for a multiwell slug test performed at a monitoring site in Lincoln County, Kansas (test initiated in well Ln-2; h_{ow} measured in well Ln-3 at a distance of 6.45 m from Ln-2; apparatus of Figure 10.5 in observation well)

TABLE 10.1
Well Construction Information for Wells Ln-2 and Ln-3

Well Designation	r_w(m)	r_c(m)	B(m)
Lincoln County Well Ln-2	0.102	0.051	6.10
Lincoln County Well Ln-3	0.071	0.025	6.10

these tests indicating that the differences shown in Figure 10.9 are primarily due to wellbore storage effects. Figure 10.10A displays results from 2 multiwell tests performed 9 days apart. In both cases, the packer-transducer arrangement of Figure 10.5 was placed immediately above the screen in the observation well. The coincidence of normalized responses at well Ln-3 when it was used as an observation well is in marked contrast to responses observed at this well when it was employed as the test well. On 6/26/91, a series of slug tests were performed using well Ln-3 as the test

TABLE 10.2
Response Data from 6/18/91 Test #2 at Lincoln County Monitoring Site

	Well Ln-2		Well Ln-3	
Time (s)	Head (m)	Normalized Head	Head (m)	Normalized Head
1.4	2.661	0.951	0.004	0.002
2.0	2.689	0.960	0.011	0.004
2.6	2.628	0.939	0.022	0.008
3.2	2.614	0.934	0.034	0.012
3.8	2.584	0.923	0.044	0.016
4.4	2.560	0.914	0.058	0.021
5.0	2.536	0.906	0.066	0.024
5.6	2.513	0.897	0.080	0.029
6.2	2.488	0.889	0.088	0.031
6.8	2.468	0.882	0.099	0.035
7.4	2.445	0.873	0.109	0.039
8.0	2.425	0.866	0.117	0.042
8.6	2.405	0.859	0.123	0.044
9.2	2.383	0.851	0.131	0.047
9.8	2.360	0.843	0.139	0.049
10.4	2.340	0.836	0.145	0.052
13.4	2.245	0.802	0.179	0.064
16.4	2.154	0.769	0.200	0.072
19.4	2.065	0.737	0.222	0.079
22.4	1.984	0.709	0.236	0.084
25.4	1.909	0.682	0.251	0.090
28.4	1.835	0.655	0.262	0.094
31.4	1.767	0.631	0.268	0.096
34.4	1.703	0.608	0.279	0.100
37.4	1.638	0.585	0.284	0.101
40.4	1.580	0.564	0.287	0.102
43.4	1.523	0.544	0.290	0.104
46.4	1.469	0.525	0.294	0.105
49.4	1.418	0.507	0.298	0.106
52.4	1.367	0.488	0.298	0.106
55.4	1.319	0.471	0.298	0.106
58.4	1.276	0.456	0.298	0.106
61.4	1.231	0.440	0.294	0.105
64.4	1.191	0.425	0.294	0.105
67.4	1.150	0.411	0.290	0.104
70.4	1.113	0.398	0.290	0.104
73.4	1.076	0.384	0.287	0.102
76.4	1.042	0.372	0.284	0.101
79.4	1.008	0.360	0.279	0.100
82.4	0.977	0.349	0.279	0.100
85.4	0.947	0.338	0.276	0.099
88.4	0.920	0.329	0.273	0.097

TABLE 10.2 (continued)
Response Data from 6/18/91 Test #2 at Lincoln County Monitoring Site

	Well Ln-2		Well Ln-3	
Time (s)	Head (m)	Normalized Head	Head (m)	Normalized Head
91.4	0.890	0.318	0.268	0.096
94.4	0.862	0.308	0.265	0.095
97.4	0.839	0.300	0.262	0.094
100.4	0.812	0.290	0.258	0.092
115.4	0.703	0.251	0.239	0.085
130.4	0.611	0.218	0.222	0.079
145.4	0.534	0.191	0.207	0.074
160.4	0.469	0.167	0.189	0.068
175.4	0.414	0.148	0.174	0.062
190.4	0.368	0.131	0.159	0.057
205.4	0.327	0.117	0.145	0.052
220.4	0.293	0.105	0.131	0.047
235.4	0.265	0.095	0.123	0.044
250.4	0.238	0.085	0.112	0.040
265.4	0.218	0.078	0.108	0.039
280.4	0.198	0.071	0.102	0.036
295.4	0.181	0.064	0.094	0.034
310.4	0.163	0.058	0.090	0.032
325.4	0.150	0.054	0.080	0.029
340.4	0.136	0.049	0.071	0.026
355.4	0.130	0.046	0.068	0.024
370.4	0.120	0.043	0.065	0.023
385.4	0.113	0.040	0.058	0.021
400.4	0.103	0.037	0.057	0.020
415.4	0.100	0.036	0.050	0.018
430.4	0.093	0.033	0.050	0.018
445.4	0.085	0.031	0.047	0.017
460.4	0.082	0.029	0.043	0.015
475.4	0.075	0.027	0.043	0.015
490.4	0.072	0.026	0.039	0.014
506.6	0.068	0.024	0.039	0.014
525.8	0.065	0.023	0.036	0.013
541.4	0.062	0.022	0.036	0.013
557.0	0.058	0.021	0.036	0.013
578.6	0.055	0.020	0.036	0.013
598.4	0.052	0.019	0.032	0.012
613.4	0.051	0.018	0.028	0.010
628.4	0.048	0.017	0.028	0.010
643.4	0.045	0.016	0.028	0.010
666.2	0.042	0.015	0.025	0.009
681.2	0.041	0.015	0.025	0.009

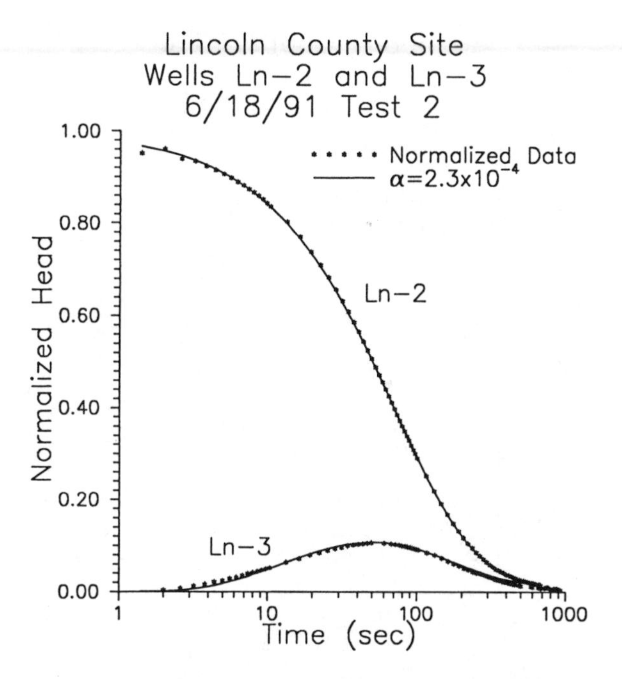

FIGURE 10.8 Normalized head ($H(t)/H_0$ and $h_{ow}(t)/H_0$) vs. logarithm of time plot for the Lincoln County multiwell slug test of Figure 10.7 and the best-fit Cooper et al. type curves.

well. Figure 10.10B very clearly shows the progressive development of a low-conductivity skin at well Ln-3 during that series of tests. However, when well Ln-3 was packed off and used as an observation well on the following day (6/27/91), Figure 10.10A indicates that this low-conductivity skin had essentially no effect on the response data. Thus, the data from these tests clearly demonstrate the advantages of using a packer in the observation well during multiwell slug tests.

The Chu and Grader Method

If the Cooper et al. method does not produce consistent α estimates from an analysis of response data at the test and observation wells or if a packer was not placed in the observation well, the method proposed by Chu and Grader (1991) should be used.

The Chu and Grader method is based on a mathematical model that can be defined as follows:

$$\frac{\partial^2 h}{\partial r^2} + \frac{1}{r}\frac{\partial h}{\partial r} + \frac{1}{r^2}\frac{\partial^2 h}{\partial \theta^2} = \frac{S_s}{K_r}\frac{\partial h}{\partial t} \tag{10.4a}$$

$$h(r,\theta,0) = 0,\ r_w < r < \infty,\ 0 \le \theta \le 2\pi \tag{10.4b}$$

$$H(0) = H_0 \tag{10.4c}$$

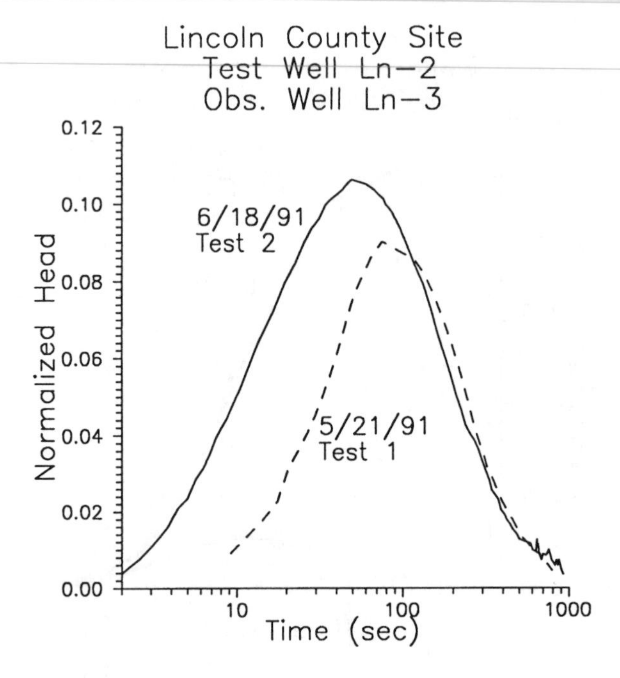

FIGURE 10.9 Normalized head ($h_{ow}(t)/H_0$) vs. logarithm of time plot for two multiwell slug tests performed at the Lincoln County monitoring site (well Ln-3 6.45 m from Ln-2; 5/21/91 test performed without packer in the observation well; 6/18/91 test performed with the apparatus of Figure 10.5 in the observation well).

$$H_{ow}(0) = 0 \tag{10.4d}$$

$$h(\infty, \theta, t) = 0,\ 0 \le \theta \le 2\pi,\ t > 0 \tag{10.4e}$$

$$H(t) = \frac{1}{2\pi}\int_0^{2\pi}\left(h(r_w, \theta_1, t) - \sigma r_w \frac{\partial h(r_w, \theta_1, t)}{\partial r}\right) d\theta_1, t > 0 \tag{10.4f}$$

$$\pi r_c^2 \frac{dH(t)}{dt} = K_r B r_w \int_0^{2\pi} \frac{\partial h(r_w, \theta_1, t)}{\partial r} d\theta_1,\ t > 0 \tag{10.4g}$$

$$H_{ow}(t) = \frac{1}{2\pi}\int_0^{2\pi}\left(h(r_{ow}, \theta_2, t) - \sigma_{ow} r_{ow} \frac{\partial h(r_{ow}, \theta_2, t)}{\partial r}\right) d\theta_2,\ t > 0 \tag{10.4h}$$

$$\pi r_{owc}^2 \frac{dH_{ow}(t)}{dt} = K_r B r_{ow} \int_0^{2\pi} \frac{\partial h(r_{ow}, \theta_2, t)}{\partial r}\, d\theta_2,\ t > 0 \tag{10.4i}$$

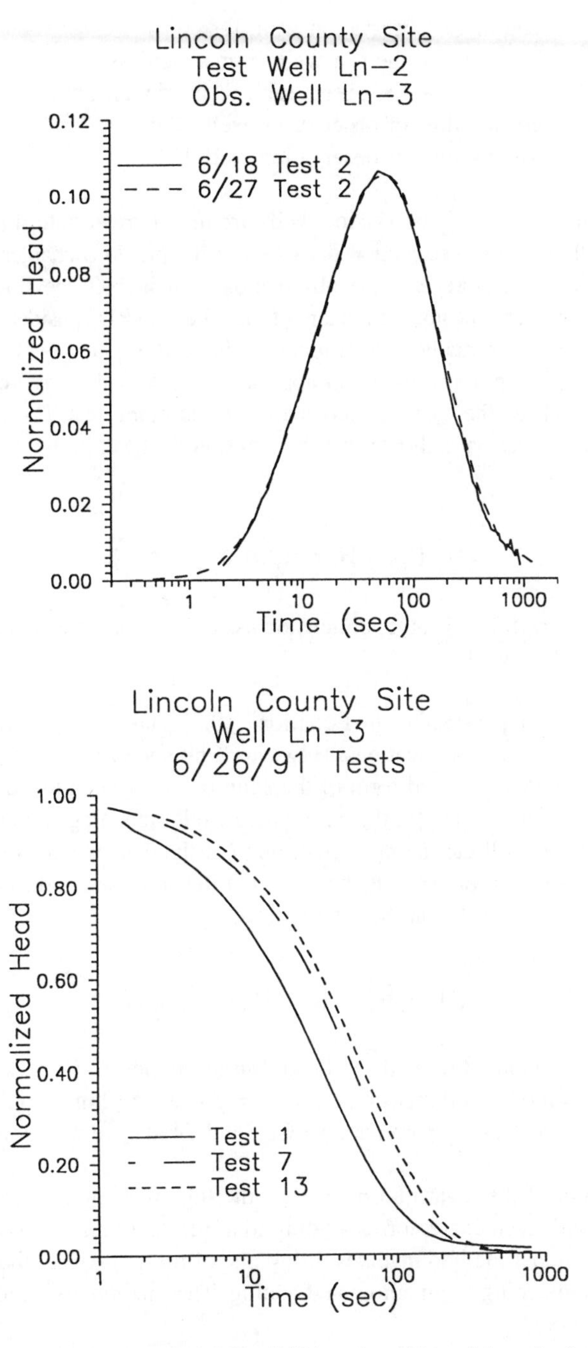

FIGURE 10.10 (A) Normalized head ($h_{ow}(t)/H_0$) vs. logarithm of time plot for two multiwell slug tests performed at the Lincoln County monitoring site (well Ln-3 6.45 m from Ln-2; both tests performed with the apparatus of Figure 10.5 in the observation well); (B) Normalized head ($H(t)/H_0$) vs. logarithm of time plot for a series of slug tests performed at well Ln-3.

where: H_{ow} = head in the observation well, [L];
σ = skin factor at the test well, [dimensionless];
σ_{ow} = skin factor at the observation well, [dimensionless];
r_{ow} = screen radius of observation well, [L];
r_{owc} = casing radius of observation well, [L].

Skin effects at the test and observation wells are incorporated into this model using the infinitely thin skin representation discussed in the previous chapter. Note that the inclusion of wellbore storage at the observation well imbues the flow field with a component of angular flow not seen in any of the other models considered in this book.

The solution to the mathematical model defined in Equations (10.4a) to (10.4i) can be obtained using a superposition approach first described by Tongpenyai and Raghavan (1981) for the case of a constant-rate pumping test. For multiwell slug tests, Chu and Grader have shown that the solution for head in the observation well can be written as:

$$H_{ow}(r_D,t)/H_0 = f(\beta,\alpha,\alpha_{ow},\sigma,\sigma_{ow}) \tag{10.5}$$

where: α_{ow} = $(r_{ow}^2 S_s B)/r_{owc}^2$, the dimensionless storage parameter for the observation well.

The number of parameters in Equation (10.5) make the general solution of limited use for practical applications. However, Chu and Grader have demonstrated that an approximate simplified form of the solution can be employed for parameter estimation in two situations: (1) the case where wellbore storage and the skin factor at the observation well can be neglected, and (2) the case where wellbore storage at the test and observation wells is the same but the skin factors differ. Under these conditions Equation (10.5) can be rewritten as:

$$(H_{ow}/H_0)*2r_D^2\alpha = H_{md} = f(\tau_{md},\gamma) \tag{10.6}$$

where: H_{md} = modified normalized head, [dimensionless];
τ_{md} = modified dimensionless time = $\beta/(\alpha^* r_D^2)$, [dimensionless];
γ = correlating parameter of Chu and Grader, [dimensionless].

Thus, a plot of the logarithm of H_{md} vs. the logarithm of τ_{md} forms a series of type curves, with each curve corresponding to a different value of γ (Figure 10.11). The Chu and Grader method involves fitting one of the γ curves to the field data via manual curve matching or an automated analog. The method essentially consists of the following six steps:

1. The logarithm of the normalized response data from the observation well is plotted vs. the logarithm of the time since test initiation;
2. The data plot is overlain by a type-curve plot prepared on graph paper of the same format (i.e., number of log cycles). The type curves are moved

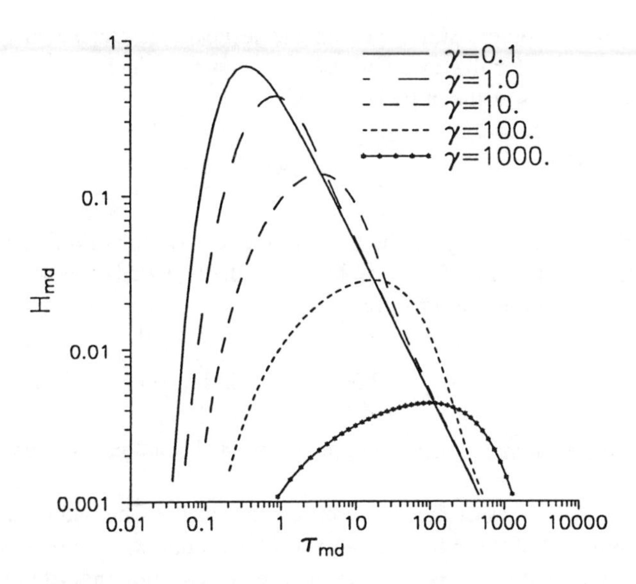

FIGURE 10.11 Logarithm of modified normalized head (H_{md}) vs. logarithm of modified dimensionless time (τ_{md}) type curves generated using the approximate method of Chu and Grader (H_{md}, τ_{md}, and γ defined in text).

parallel to the x and y axes of the data plot until one of the γ curves approximately matches the plot of the field data. Unlike the method of Cooper et al., the y axes are shifted with respect to one another during this process;

3. Match points are selected from each plot. For convenience's sake, H_{md} and τ_{md} are set to 0.01 and 1.0, respectively, and the normalized head and time corresponding to these values are read from the data plot;
4. An estimate for the specific storage is calculated from the definition of H_{md}:

$$S_s = \frac{\left(\dfrac{0.01}{H_{ow}/H_0}\right) r_c^2}{2Br_D^2 r_w^2} \tag{10.7a}$$

where the terms in the parenthesis are the match points from the y axis;

5. An estimate for the radial component of hydraulic conductivity is calculated from the definition of τ_{md}:

$$K_r = \left(\frac{1.0}{t}\right) S_s r_w^2 r_D^2 \tag{10.7b}$$

where the terms in the parenthesis are the match points from the x axis;

6. An estimate for the skin factor at the test well is calculated from the definition of γ for the case when wellbore storage and the skin factor at the observation well can be neglected:

$$\sigma = \alpha\gamma_{cal} r_D^2 + 0.5\ln(2\alpha) \tag{10.7c}$$

When wellbore storage is the same at the test and observation wells, an estimate for the sum of the skin factors at the two wells can be calculated from the definition of γ for that case:

$$(\sigma + \sigma_{ow}) = 2.5\alpha\gamma_{cal} r_D^2 + 1.25\ln(2\alpha) \tag{10.7d}$$

For both situations, γ_{cal} is the γ value determined from the type curve match.

There are three issues of practical importance for the Chu and Grader method. The first issue is that of the significance of wellbore storage at the observation well. The results of the field tests described in the previous section indicate that wellbore storage at the observation well can often be ignored when a packer is placed in the observation well. The solution of Chu and Grader can be used to examine the general conditions under which the effects of wellbore storage can be neglected. Figure 10.12 displays the results of a series of simulations that demonstrate the convergence of the Chu and Grader solution on that of Cooper et al. when a packer is placed in the observation well and skin effects are assumed negligible. As described in Chapter 7, use of a packer in the observation well produces a very small effective casing radius and, thus, a very large α_{ow} (>0.01). At that α_{ow} value, the amount of water moving between the well and the formation for head equilibration becomes extremely small, effectively eliminating the impact of wellbore storage effects on the response data.

The second issue of practical importance is that of the impact of a well skin at the observation well. The field tests described in the previous section demonstrated that the effect of a well skin at the observation well is often quite small when a packer is used in the observation well, because only a very small amount of water needs to move between the observation well and the formation for head equilibration. Figure 10.13A displays the results of a series of simulations performed with the solution of Chu and Grader to further demonstrate the impact of a low-conductivity skin when a packer is used in the observation well. Since the difference between σ_{ow} curves becomes much smaller with increases in r_D, these results indicate that the effects of a well skin at the observation well can be neglected when a packer is used. Figure 10.13B shows analogous results for the case of equal wellbore storage at the test and observation wells, demonstrating the critical role of a packer in determining the significance of σ_{ow}. Although Figure 10.13C indicates that the effect of σ_{ow} decreases with increases in r_D, σ_{ow} will have a significant effect on the response data for the r_D commonly used for shallow groundwater applications when a packer is not placed in the observation well.

Third, when neither of the conditions assumed for the two variants of the γ type curve method are appropriate, parameter estimation must be based on Equation (10.5).

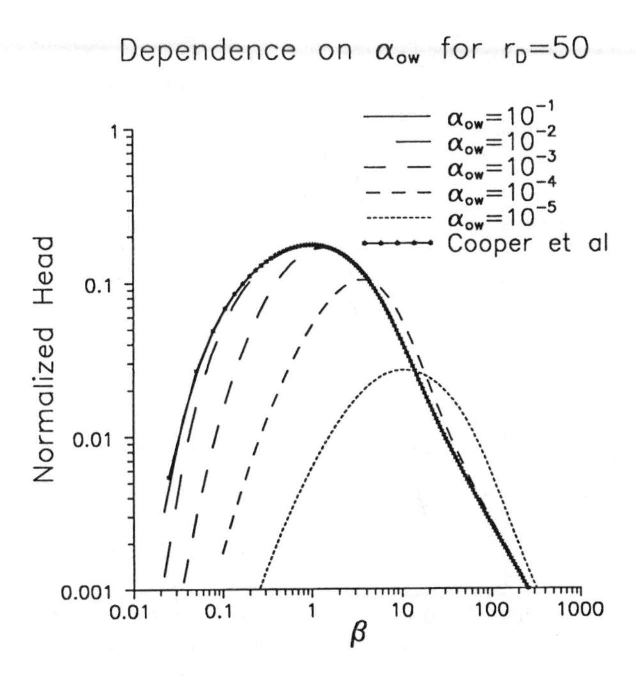

FIGURE 10.12 Logarithm of normalized head ($h_{ow}(t)/H_0$) vs. logarithm of β type curves generated with the solution of Chu and Grader (α_{ow} and r_D defined in text).

In that case, the number of parameters greatly complicates the analysis procedure, making the additional information potentially available from response data at an observation well difficult to exploit. Thus, it is strongly recommended that packers be used in observation wells to maximize the information obtainable from multiwell tests. For best results, the packer should be placed just above the top of the screen.

The Lincoln County slug test discussed in the previous section can also be used to illustrate an application of the Chu and Grader method. Since a packer was placed in the observation well, the γ definition used in Equation (10.7c) and the type curves of Figure 10.11 are employed in the analysis. Figure 10.14 shows the match that is obtained with the $\gamma = 10$ type curve. Substitution of the match points found from the analysis into Equations (10.7a) to (10.7c) yields estimates of 6.6×10^{-6} m^{-1}, 1.57 m/day, and 2.42 for S_s, K_r, and σ, respectively. Note that the K_r estimate is within 48% of that obtained from the multiwell variant of the Cooper et al. method, while the S_s estimate is within 22% of that obtained from the Cooper et al. method. These results were obtained with a manual matching approach only using a few type curves. An automated approach, which would allow type curves of any γ to be used, would undoubtedly have produced estimates in better agreement with those found with the Cooper et al. method.

Methods for Partially Penetrating Wells

Very few applications of multiwell slug tests have been reported in the literature for tests initiated in partially penetrating wells. The few reported cases have used one of three approaches to analyze the response data at the observation well: (1) the

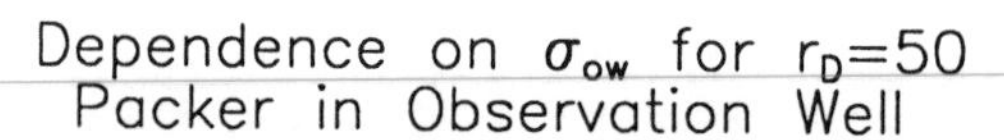

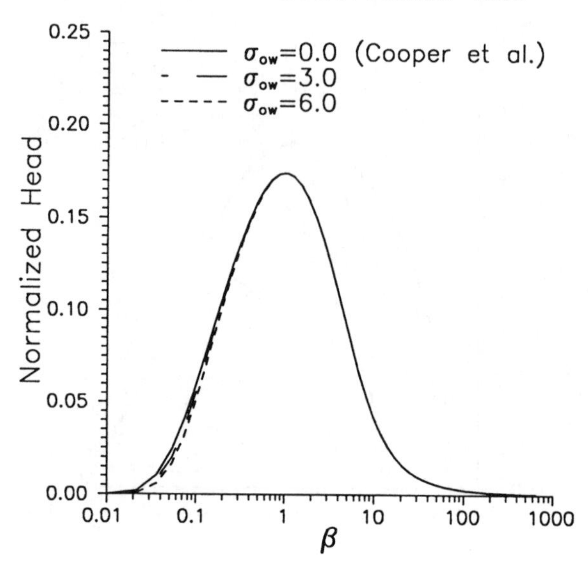

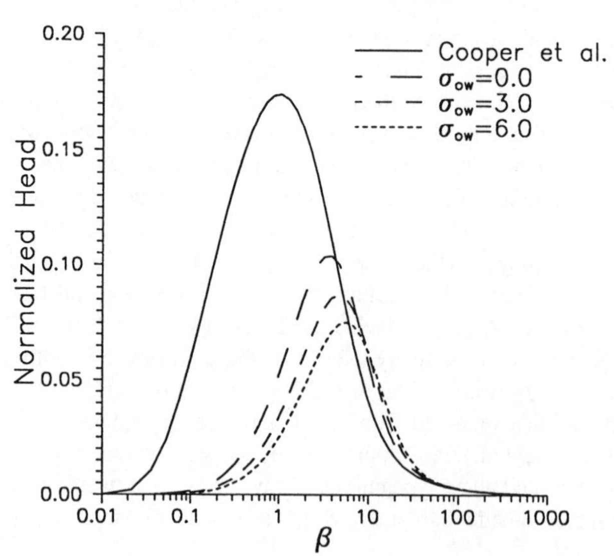

FIGURE 10.13 Normalized head ($h_{ow}(t)/H_0$) vs. logarithm of β type curves generated with the solution of Chu and Grader: (A) Packer in the observation well, r_D = 50; (B) No packer in the observation well, r_D = 50; (C) No packer in the observation well, r_D = 500 (σ_{ow} and r_D defined in text; solution of Cooper et al. provided for comparison).

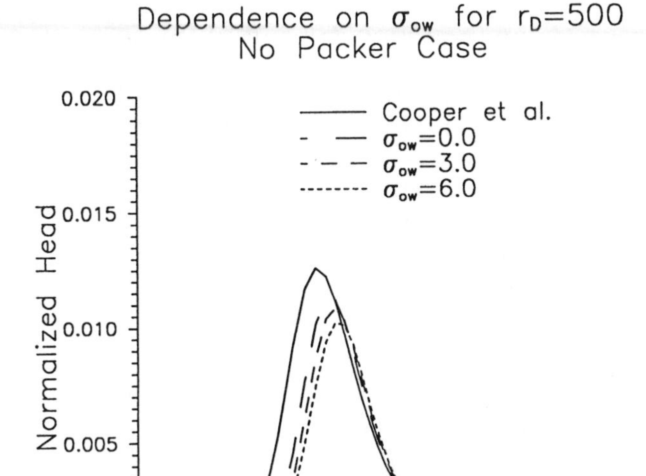

FIGURE 10.13 (continued)

multiwell variant of the Cooper et al. method, (2) the multiwell variant of the KGS model, and (3) the method of Spane and co-workers. Each of these methods will briefly be described in the following paragraphs.

The multiwell variant of the Cooper et al. method is based on the mathematical model defined by Equations (5.1a) to (5.1f). When applied to the case of a partially penetrating well, the quantity, B, formation thickness, is replaced by b, the effective screen length. The vertical component of slug-induced flow is ignored and the flow is assumed to be purely radial in nature. The steps employed for the analysis are exactly the same as those used for a fully penetrating well. As with single-well tests, the appropriateness of the parameter estimates will depend on the ψ parameter. In cases where the vertical anisotropy ratio is near one, Butler and McElwee (1996) have shown that use of the Cooper et al. method can lead to a significant overestimation of K_r and S_s. Thus, unless there is strong evidence to indicate that ψ is quite small, i.e., the vertical component of hydraulic conductivity is significantly less than the horizontal component and/or the test well is of a very large aspect ratio, the multiwell variant of the Cooper et al. method should not be used to analyze response data from a slug test initiated in a partially penetrating well.

The multiwell variant of the KGS model is based on the mathematical model defined in Equations (5.24a) to (5.24g) for confined condition and Equations (6.7a) to (6.7h) for unconfined conditions. In both cases, wellbore storage and skin effects at the observation well are neglected; so, a packer must be used in the observation well. The implementation of the approach is similar to the single-well form of the KGS model. However, unlike the single-well case, response data at an observation well are quite sensitive to both α and ψ (Figures 10.15A and 10.15B). This dependence greatly

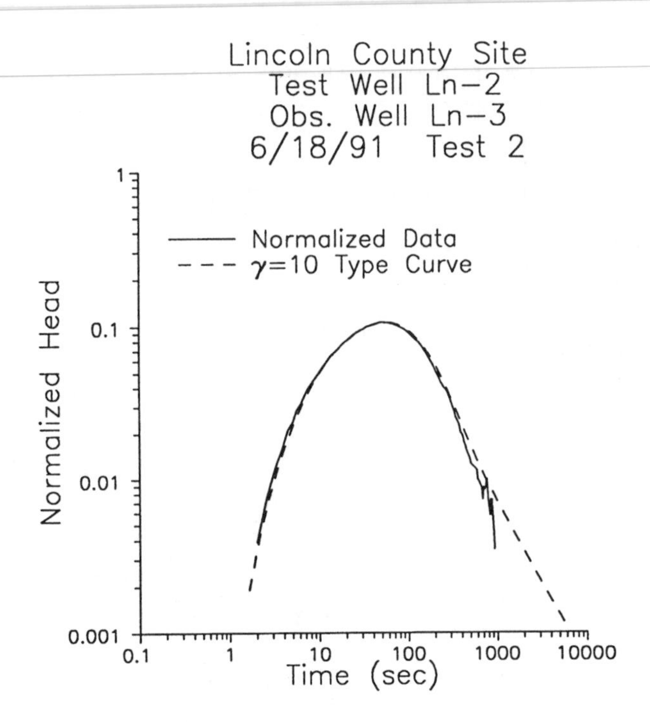

FIGURE 10.14 Logarithm of normalized head ($h_{ow}(t)/H_0$) vs. logarithm of time plot for the Lincoln County multiwell slug test of Figure 10.7 and the best-fit Chu and Grader type curve (γ defined in text).

complicates the manual type-curve matching procedure as for each α value there is a suite of ψ type curves (e.g., Figure 10.15B) for a well at a given r_D. Thus, automated matching procedures are the only realistic option. Although more complicated, this approach can potentially provide very valuable information about the vertical anisotropy ratio and specific storage.

Spane and co-workers (Spane, 1996; Spane et al., 1996) describe an approach for the analysis of multiwell slug tests that is based on an extension of the solution of Neuman (1975) for pumping-induced drawdown in an unconfined formation and the general relationship between pumping-test and slug-test solutions discussed by Peres et al. (1989). This method is actually quite similar to that based on the unconfined form of the KGS model. The major difference between the two approaches is that the position of the water table can change in the solution of Neuman (1975) so it is theoretically possible to get information about specific yield with the method of Spane et al. However, the magnitude of normalized heads at which information about specific yield can be obtained is so small that it is very difficult to exploit this advantage in most shallow groundwater applications. Thus, the Spane et al. method should yield information quite similar to that obtained with the unconfined form of the KGS model. Spane et al. (1996) discuss two field examples to demonstrate the potential of the approach, one of which involves a multilevel observation well. The use of a multilevel well enables response data to

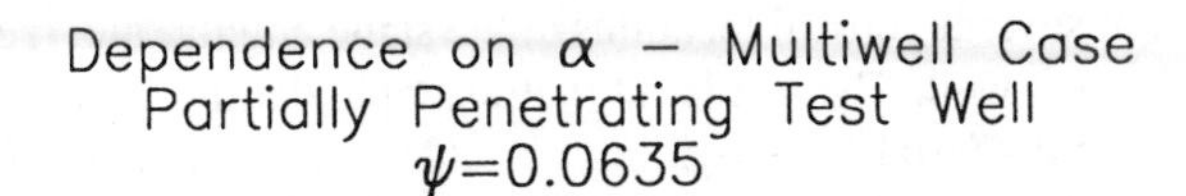

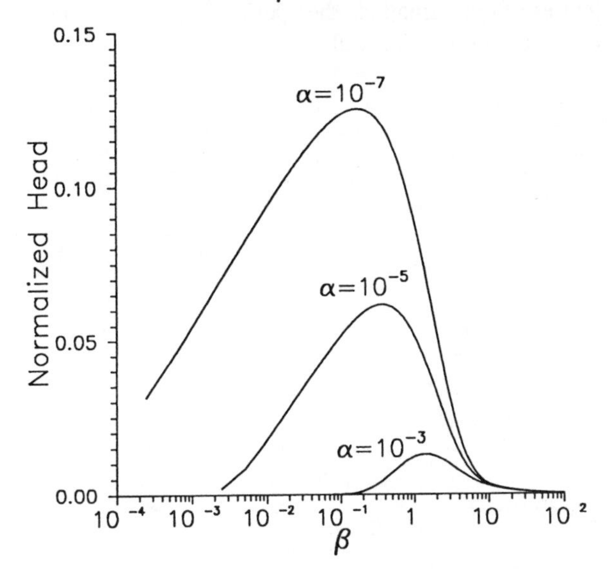

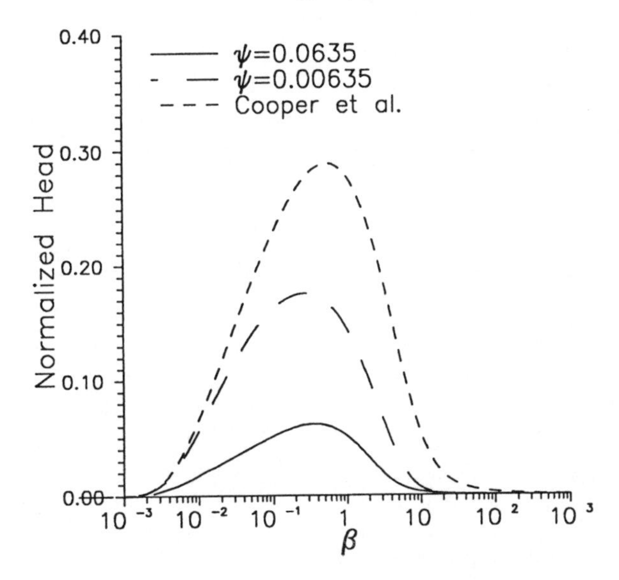

FIGURE 10.15 (A) Normalized head ($h_{ow}(t)/H_0$) vs. logarithm of β type curves generated with the KGS model for a specific ψ; (B) Normalized head ($h_{ow}(t)/H_0$) vs. logarithm of β type curves generated with the KGS model for a specific α ($r_D = 50$ in both cases).

be collected at several vertical locations at the same angular and radial position with respect to the test well. The results of that example demonstrate that multiwell slug tests have great potential for providing valuable information about the transmissive and storage properties of the formation when performed in the vicinity of a piezometer nest or a multilevel observation well.

11 The Analysis of Slug Tests — Additional Issues

CHAPTER OVERVIEW

All of the analysis methods discussed in the previous chapters have been based on mathematical models that represent the formation as a porous, naturally homogeneous medium. This representation, however, may not always be appropriate. In this chapter, additional considerations for slug tests performed in fractured and naturally heterogeneous formations are reviewed. A brief discussion is also presented on the use of slug tests for assessing conditions at prospective pumping-test observation wells.

SLUG TESTS IN FRACTURED FORMATIONS

Conventional analysis methods are based on the assumption that the test interval can be conceptualized as a porous medium. However, in cases where fractures serve as major conduits for flow, the viability of this representation may be in question. The porous-medium conceptualization is most appropriate in fractured formations where the density of conductive fractures is very high and where there is either virtually no fluid exchange between the matrix and the fractures or the rate of exchange is extremely rapid. When these conditions are not met, alternate conceptualizations must be used. One alternative is the discrete-fracture model (Wang et al., 1977; Karasaki et al., 1988), which is appropriate for conditions when the density of conductive fractures is low and there is little exchange between fractures and the matrix. A second alternative is the double-porosity model (Moench, 1984), the key features of which are that the density of conductive fractures is moderately high and there is some fluid exchange between fractures and the matrix. Techniques for the analysis of response data have been developed for both of these classes of models and will be briefly discussed in the following two sections.

DISCRETE FRACTURE MODELS

The most thorough assessment of slug tests in media that can be represented by the discrete-fracture model is presented by Wang et al. (1977). A major conclusion of their work is that the existence of one or a series of discrete conductive fractures will be difficult to recognize with slug tests, as the response data may be quite similar to what would be seen in a porous medium. However, if the location of discrete fractures in a borehole can be identified from geophysical logging or other means, individual fractures can be isolated and tested using a conventional straddle-packer arrangement. Standard models for slug tests in confined formations can then be used

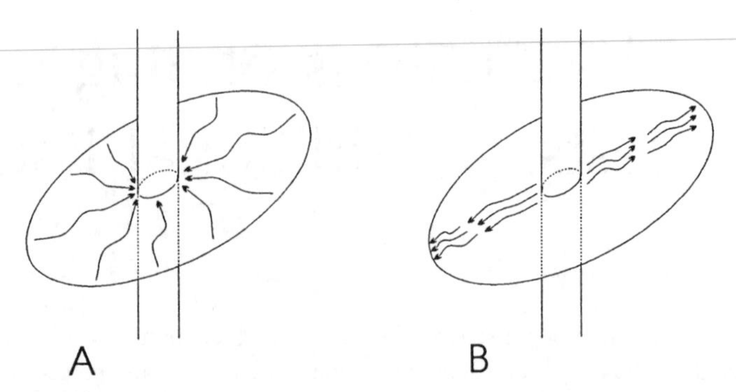

FIGURE 11.1 Schematic of two possible conditions of flow in a single fracture: (A) Flow in the entire plane of the fracture; and (B) Flow restricted to a narrow channel within the fracture plane. (After Karasaki et al., 1988).

to analyze the response data and estimate the aperture (width) of the fracture. The hydraulic conductivity estimate obtained from the analysis can be related to the aperture of the fracture using a relationship derived from parallel plate models of fracture flow (Wang et al., 1977):

$$K_r = \frac{b_f^2 g}{12\nu} \tag{11.1}$$

where: b_f = fracture aperture, [L].

The work of Wang et al. (1977) is based on the assumption that flow can occur through the entire plane of the fracture (e.g., Figure 11.1A). In many situations, however, the flow may be restricted to relatively narrow channels within the fracture plane (e.g., Figure 11.1B). Karasaki et al. (1988) introduce the linear-flow model as a means to analyze slug tests in certain cases of channelized flow. Unlike radial flow models, where the cross-sectional area through which the slug-induced flow occurs increases as a linear function of distance from the test well, the linear-flow model represents conditions where the cross-sectional area for flow does not change with distance from the well, such as might be produced by channelized flow within a fracture plane of an arbitrary orientation (e.g., Figure 11.1B). Karasaki et al. (1988) develop an analytical solution for this model that can be written as:

$$H(t)/H_0 = f\left(\alpha_{lf}, \beta_{lf}\right) \tag{11.2}$$

where: $\alpha_{lf} = (Ar_w^2 S_s)/r_c^2$;
$\beta_{lf} = (AKt)/r_c^2)$;
A = (cross-sectional area for flow)/(πr_w).

The form of Equation (11.2) reveals that it will be extremely difficult to obtain a reasonable estimate of hydraulic conductivity from a slug test in a formation in

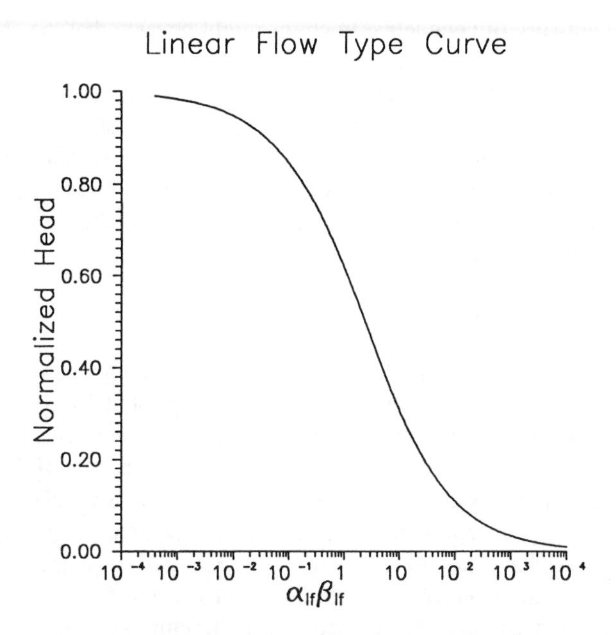

FIGURE 11.2 Normalized head ($H(t)/H_0$, where H(t) is deviation from static and H_0 is magnitude of the initial displacement) vs. logarithm of $\alpha_{lf}\beta_{lf}$ type curve generated with the linear flow model of Karasaki et al. (α_{lf} and β_{lf} defined in text).

which the linear-flow model is an appropriate representation of the governing physics. Since α_{lf} and β_{lf} appear together as a product in Equation (11.2), a very good estimate of α_{lf} (i.e., A, r_w, and S_s) must be available to have any hope of obtaining a reasonable estimate of hydraulic conductivity. Estimation of the cross-sectional area for flow (A) will be particularly difficult in field settings. Given these limitations, the linear-flow model is of little use for the estimation of hydraulic conductivity. Instead, the primary purpose of this model should be to assess the significance of linear flow during a slug test in fractured media.

Figure 11.2 is a graph of the type curve that results from Equation (11.2). This type curve is in the form of a plot of normalized head vs. the logarithm of dimensionless time, where dimensionless time is defined here as the product $\alpha_{lf}\beta_{lf}$. As shown in Figure 11.2, the diagnostic feature of the linear-flow model is a response plot that is much broader than that seen in most radial-flow cases (e.g., Figure 5.2). An α value considerably greater than 1.0 would be required in the radial-flow model to match the pattern shown on Figure 11.2 (e.g., Figure 7.6). Since such a large α is extremely rare in slug tests performed in open wells, a very broad response plot should be considered strong evidence for the presence of linear flow. Unfortunately, other than identifying the importance of channelized flow for that particular test, relatively little can be gained from a slug test in a formation in which a linear-flow model represents the governing physics.

The linear-flow model is appropriate when the cross-sectional area through which flow occurs does not change with distance from the well. In some cases, however, the network of conductive fractures might be such that the cross-sectional area does change with distance from the well, but in a manner different from what would be predicted with a conventional radial- or cylindrical-flow model. In this case, a fractional-dimension representation may be the most appropriate approach for conceptualization of the flow regime. Barker (1988) derives an analytical solution for slug tests in systems where such a representation would be appropriate. Unfortunately, however, uncertainty regarding the dimension of the flow regime makes the solution rather difficult to apply in practice.

Double Porosity Models

The double-porosity model is probably the most common approach for representing flow in fractured systems. Barker and Black (1983) present an analytical solution for a slug test performed in a fully penetrating well in a fractured formation that is conceptualized as a series of slabs separated by equally spaced horizontal fractures. Ramey and co-workers (Mateen and Ramey, 1984; Grader and Ramey, 1988) extend the approach to more general fracture geometries, while Dougherty and Babu (1984) present a solution for a test in a partially penetrating well. Although the various solutions differ in the manner in which they represent the fluid exchange between fractures and the matrix and the geometry of the fracture system, these differences are of little practical significance. Using the simplest representation of the exchange between fractures and the matrix, the analytical solution for the case of a slug test in a fully penetrating well can be written as:

$$H(t)/H_0 = f(\beta, \alpha, \omega, \lambda) \tag{11.3}$$

where: ω = double-porosity storage parameter, ratio of specific storage of fractures over specific storage of both matrix and fractures, [dimensionless];
λ = interporosity flow parameter, characterizes ease of fluid movement between matrix and fractures, [dimensionless].

The α and β parameters are as defined previously for a porous medium (e.g., Equation [5.2]). The extension of Equation (11.3) to a partially penetrating well is analogous to the approach used for a porous medium (Dougherty and Babu, 1984).

Regardless of the manner in which the various models represent the fluid exchange between fractures and the matrix and the geometry of the fracture system, the resulting solutions will always be in a form similar to Equation (11.3) in that two new parameters are added to the parameter set used for conventional porous-media applications. The critical practical issue for double-porosity models is that of uniqueness, i.e., do these additional two parameters modify the slug-induced responses to such an extent that the existence of a double-porosity flow regime is evident from the response data? Unfortunately, the answer to this question is no, at least not for the normalized heads commonly measured in shallow groundwater investigations. Figure 11.3A is a comparison between responses simulated for a slug test in a fully penetrating well using the Cooper et al. model (labelled single porosity)

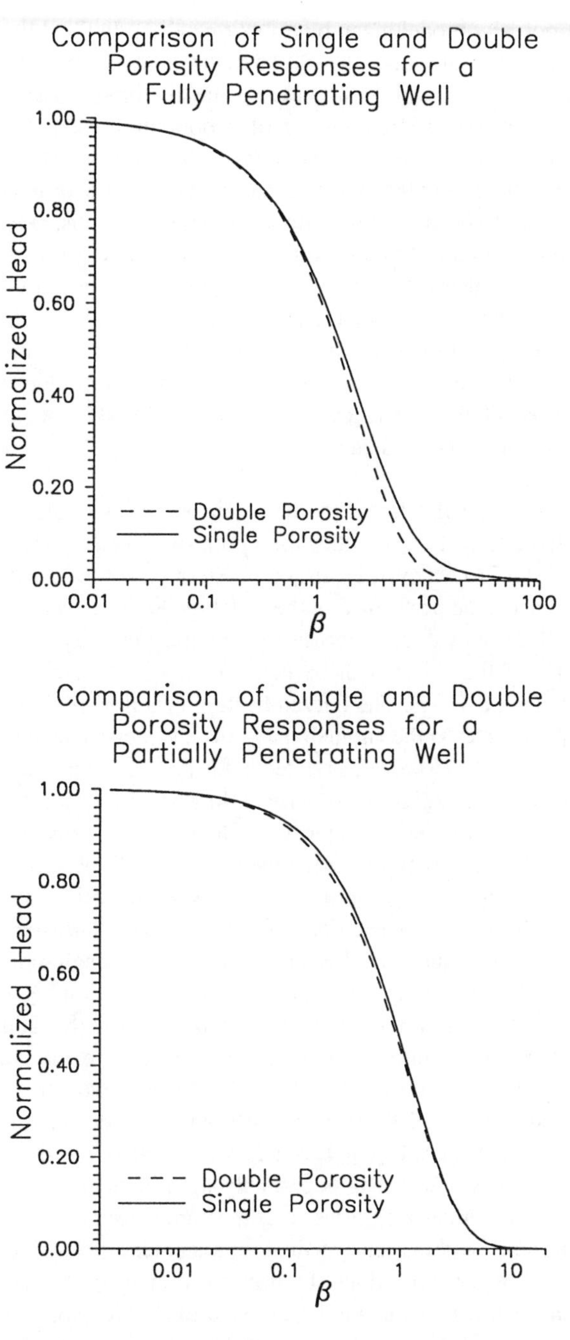

FIGURE 11.3 (A) Normalized head ($H(t)/H_0$) vs. logarithm of β plots generated with the Cooper et al. model (labelled single porosity) and its double-porosity extension; (B) Normalized head ($H(t)/H_0$) vs. logarithm of β plots generated with the KGS model (labelled single porosity) and its double-porosity extension.

and the double-porosity model given by Equation (11.3). In this hypothetical example, the ω parameter is quite small to reflect a matrix that has a much higher porosity than the fracture system, and to accentuate the double-porosity aspects of the responses. Note that the primary impact of a double-porosity flow regime is to steepen the response plot relative to the single-porosity case. However, as Barker and Black (1983) have pointed out, this steepening of the response plot can be mimicked in the single-porosity case with a very small α. Thus, response data from a slug test performed in a fully penetrating well in a double-porosity formation can be fit extremely well using the model of Cooper et al. The only indication that double-porosity processes are of significance will be an α estimate that is lower than expected. However, given the uncertainty about α that exists for most tests and the fact that a smaller-than-expected α estimate can also be a product of a significant component of vertical flow or a low-permeability well skin, it may be virtually impossible to recognize the existence of a double-porosity flow regime from the α estimate.

This situation is best illustrated by the extensive series of slug tests performed as part of site-characterization activities for a proposed nuclear waste repository, the Waste Isolation Pilot Project (WIPP), in southeastern New Mexico. Although a large number of tests have been performed in the Culebra Dolomite, a unit that is strongly suspected to behave as a double-porosity formation with respect to groundwater flow, virtually all of those tests can be fit extremely well using the Cooper et al. model (e.g., Beauheim, 1987). The only indication of a flow regime with a possible double-porosity character is the anomalously low α values calculated for some tests ($1 \times 10^{-12} \leq \alpha \leq 1 \times 10^{-7}$). However, as stated in the previous paragraph, anomalously low α values can also be explained by vertical flow or a low-permeability well skin; so, such values are by no means diagnostic of a double-porosity flow regime.

Given that a double-porosity flow regime cannot easily be recognized from any characteristic of test responses, the decision of when to utilize a double-porosity model for the analysis of response data can be difficult. Thus, the issue of how a hydraulic conductivity estimate is affected by the use of a single-porosity model for the analysis of response data from a double-porosity formation becomes important. Barker and Black (1983) show that for a fully penetrating well the failure to recognize the existence of a double-porosity flow regime can lead to an overestimation of hydraulic conductivity. According to Black (1985), however, that overestimation should not exceed a factor of three, and will usually be less than a factor of two. Both Barker and Black (1983) and Black (1985) point out that use of a double-porosity model to analyze test data is probably not practical because of the nonuniqueness of the theoretical responses, i.e., the same responses can be generated using many different combinations of model parameters. Thus, these authors recommend that the Cooper et al. model be utilized to analyze the response data and that one be aware that use of this model can lead to an overestimation of hydraulic conductivity in fractured systems. Figures presented in both articles can be used to assess the conditions under which the overestimation of hydraulic conductivity is apt to be of practical concern.

Since most slug tests are performed in wells that are screened over a limited portion of a formation, the above discussion must be extended to the case of a partially penetrating well. Figure 11.3B is a comparison between simulated single-porosity (Hyder et al., 1994) and double-porosity responses for hypothetical slug tests in partially penetrating wells. The same parameters are used as in the simulations of Figure 11.3A, with the exception that the well is assumed to be screened over a short vertical interval at the center of a thick formation ($\psi = 0.0635$). It is apparent from Figure 11.3B that the differences between single- and double-porosity responses are quite small for a slug test performed in a partially penetrating well. For wells of small to moderate aspect ratios (moderate to large ψ), the significant component of vertical flow largely masks the impact of double-porosity mechanisms for the normalized head ranges measured in conventional slug tests. Thus, under the conditions most commonly faced in field applications, the presence of a double-porosity flow regime will introduce very little error into the hydraulic conductivity estimate obtained with a single-porosity model.

As emphasized in the preceding paragraphs, the existence of a double-porosity flow regime will be virtually impossible to detect with the normalized heads commonly measured in conventional shallow groundwater investigations. Petroleum engineers, however, have found that some indication of double-porosity behavior may be evident if response data can be collected in the very late stages of a test (i.e., normalized heads much less than 0.01). The clarity of that indication will depend on the mechanisms governing the exchange of fluid between the matrix and the fractures (Grader and Ramey, 1988). Double-porosity behavior is most easily detected when there is a low-permeability coating on the fracture walls (e.g., Moench, 1984), thereby slowing the exchange between the fractures and the matrix. Even with a fracture coating, however, the indication of double-porosity behavior will be restricted to normalized heads well below 0.01, a range for which it can be quite difficult to obtain relatively noise-free data in tests performed as part of shallow groundwater investigations.

SLUG TESTS IN NATURALLY HETEROGENEOUS FORMATIONS

All of the analysis methods discussed in this book have been based on the assumption that the formation is homogeneous under natural conditions, i.e., any variations in the transmissive and storage properties of the formation are a product of mechanisms related to well drilling/development and subsequent deterioration. Clearly, however, this idealized representation of the formation is not in keeping with reality. Natural systems are actually characterized by a considerable degree of variability in their hydraulic properties. In this section, the limited work that has been reported on slug tests in naturally heterogeneous formations is briefly summarized.

Data collected at a variety of scales from a large number of sites have shown that the rate of variation in the hydraulic properties of a formation is greatest in the direction perpendicular to bedding planes (e.g., Smith, 1981; Sudicky, 1986; Hess et al., 1992). Thus, an often reasonable representation of a formation for the purposes of a slug test is as a series of horizontal layers in which flow properties are assumed to vary between, but not within, individual layers. Karasaki et al. (1988) were among the first to rigorously examine slug-induced flow in this configuration. They developed an analytical solution for the case of a fully penetrating well in a two-layered system in which the vertical anisotropy ratio of each layer is so small that no flow will occur between layers. Their results show that it is practically impossible to distinguish a slug test in a layered system from that in a homogeneous formation. They also demonstrated that the hydraulic conductivity estimate obtained from such a test would be the thickness-weighted arithmetic average of the radial component of hydraulic conductivity (K_r) of each layer. Their results, however, are only strictly valid for the analytically tractable condition of no vertical flow between layers. Butler et al. (1994), however, employed a numerical model to examine the more general case of unrestrained vertical flow. They found that the results of Karasaki et al. are appropriate for all fully penetrating wells, regardless of the degree of vertical flow between layers or the number of layers. Thus, slug tests performed in fully penetrating wells in layered units will provide no information about vertical variations in hydraulic properties when measurements are taken only at the test well. In all cases, the estimated conductivity will be a thickness-weighted arithmetic average of the K_r of the individual layers.

Butler et al. (1994) emphasize that these findings only pertain to tests performed in fully penetrating wells. In the case of partially penetrating wells, the appropriateness of the thickness-weighted arithmetic average will depend on the vertical anisotropy ratio and the aspect ratio, the two factors encapsulated in the ψ parameter. Butler et al. (1994) found that for ψ values much less than approximately 0.005, the hydraulic conductivity estimate obtained from a slug test is approximately equal to the thickness-weighted arithmetic average of the K_r of the layers intersecting the test interval. However, at larger ψ values, a significant component of vertical flow will cause the layers adjoining the test interval to affect the conductivity estimate. Thus, a simple average of the conductivities of the layers intersecting the test interval will no longer be appropriate. A general expression for the average conductivity obtained from a slug test in a partially penetrating well in a layered formation has apparently not yet been developed.

Slug tests are often performed in a multilevel format in which a system of straddle packers is used to test discrete intervals of the portion of the formation intersecting the open (screened) section of the well. The primary purpose of tests in this configuration is to assess the presence of thin high-conductivity zones that may act as preferential flow paths. Butler et al. (1994) have shown that both high- and low-conductivity skins can have a large impact on parameter estimates obtained from a program of multilevel slug tests. In particular, a high-conductivity skin (e.g., an artificial filter pack) can act as a conduit for vertical flow, allowing portions of the formation outside of that isolated by the straddle packers to have a significant effect on test responses. This phenomenon is of most concern in test intervals of small aspect ratios in conjunction with thick filter packs. Thus, artificial filter packs are best not used in wells at which a program of multilevel slug tests is to be performed.

The results reported in the previous paragraphs were found for a formation conceptualized as a series of homogeneous layers of infinite extent. A more realistic representation of natural systems would be a model in which hydraulic properties vary in both the vertical and horizontal directions. Harvey (1992) presents the results of a numerical modeling study of slug tests in wells fully penetrating a layered aquifer in which hydraulic conductivity varied both between and within layers. The rate of variation between layers was much greater than that within layers in keeping with the results of most field investigations (e.g., Smith, 1981; Sudicky, 1986; Hess et al., 1992). Conductivity variations were incorporated into the model using a stochastic process representation (e.g., Dagan, 1989; Gelhar, 1993). This study found that the hydraulic conductivity estimate obtained from an analysis with the Cooper et al. method in a heterogeneous formation can be approximated by an average of the small-scale conductivities. Although the exact form of this average will depend on the characteristics of the particular stochastic process used to represent the conductivity variations, these results indicate that the average will lie somewhere between the arithmetic and geometric mean of the small-scale conductivities. Further work, however, is needed to assess the generality of these results for tests in partially penetrating wells and formations in which alternate definitions of the stochastic process would be more appropriate. Although knowledge of the form of this average could be very helpful in attempts to combine estimates from slug tests with those found by other means, the inherent uncertainty in parameter estimates obtained from slug tests, arising from the well-construction parameters and other factors, will greatly limit the practical utility of such information.

If a large number of wells are available, slug tests can potentially provide very useful information about spatial variations in hydraulic conductivity at a site. Unfortunately, at most sites, only a few wells are placed within each unit of interest, thereby limiting the information that can be obtained about hydraulic conductivity variations at that site. If a pumping test has been performed, Butler (1990) suggests that the conductivity estimates obtained from a series of slug tests be compared to estimates obtained at the same wells from an analysis of the pumping-induced drawdown with the Cooper-Jacob method. Since the conductivity estimate obtained from the Cooper-Jacob method will be insensitive to the portions of the formation in the immediate vicinity of the observation well (Butler, 1988; Butler, 1990), close agreement between slug-test and pumping-test estimates should be considered strong evidence that the formation can be represented as a homogeneous unit at the slug test or larger scale.

ASSESSMENT OF PROSPECTIVE OBSERVATION WELLS

An underutilized application of slug tests is as a means of assessing the appropriateness of a particular well for use as an observation well during a pumping test. Black and Kipp (1977a, b) have shown how wellbore storage and skin effects at the observation well can affect parameter estimates obtained from pumping tests.

Although the Cooper-Jacob method can remove the impact of these mechanisms from the conductivity estimate, this technique may not be applicable for short tests or systems with significant leakage. In those conditions, type curve based methods, which can be significantly influenced by wellbore storage and skin effects at the observation well, must be used. Thus, a program of slug tests should be performed at prospective observation wells prior to a pumping test. The information gained from the slug tests can be used to both assist in the analysis of the pumping-induced drawdown, as described by Black and Kipp (1977a,b), as well as to develop some understanding of the scale of hydraulic conductivity variations within a particular unit following the approach proposed by Butler (1990). Note that the modified techniques for the analysis of pumping-induced drawdown proposed by Black and Kipp will not be necessary if a packer is used in the observation well. As described in Chapter 10, the use of a packer in an observation well will greatly diminish the impact of wellbore storage and skin effects on the head changes measured at that well. These techniques will also not be necessary if the observation well is at a relatively large distance from the pumping well (e.g., Tongpenyai and Raghavan, 1981).

12 The Analysis of Slug Tests — Guidelines

CHAPTER OVERVIEW

In the preceding seven chapters, the major methods for the analysis of response data were described. The purpose of this chapter is to summarize this material in the form of a series of practical analysis guidelines. Individual sets of guidelines will be proposed for each of the common hydrogeologic settings faced in shallow groundwater investigations.

ANALYSIS GUIDELINES

The first phase of the analysis of response data from a series of slug tests is to select a general class of analysis methods. Figure 12.1 is a decision tree outlining the major steps in this process. The first task in Figure 12.1 is to assess the appropriateness of conventional theory for that set of tests. As discussed in Chapter 2, the coincidence of normalized data plots from a series of tests at the same well is a convincing demonstration that conventional linear theory is applicable for those tests. If linear theory is applicable, the remaining steps are directed at assessing which hydrogeologic setting is most appropriate for that particular series of tests. As shown in the figure, analysis guidelines are provided here for the most common conditions faced in shallow groundwater investigations.

If the conventional linear theory does not appear appropriate for a particular series of tests, the underlying mechanisms responsible for this situation should be ascertained prior to the actual analysis. A reproducible dependence on H_0 is either a product of non-Darcian flow or of changes in the effective screen length. A dependence on flow direction similar to that displayed in Figure 2.8 is strong evidence that the response data are being affected by changes in the effective screen length during the test. A nonreproducible dependence on H_0 is an expression of a dynamic skin. Unfortunately, relatively little can be gained from an analysis of response data affected by a dynamic skin. In this case, data from the test that appears to have been least impacted by dynamic skin effects should be used in the analysis. However, the resulting conductivity estimate should only be considered a very rough approximation of the conductivity of the formation.

In the following sections, a series of practical analysis guidelines will be presented for each of the hydrogeologic settings commonly faced in shallow groundwater investigations. Unlike the guidelines presented in Chapters 2 through 4, these analysis guidelines will primarily be in the form of flow charts. Figure 12.2 is an example of such a flow chart. Notice that the terminus of any series of activities is one or more parameter estimates. The relative quality of a particular estimate is

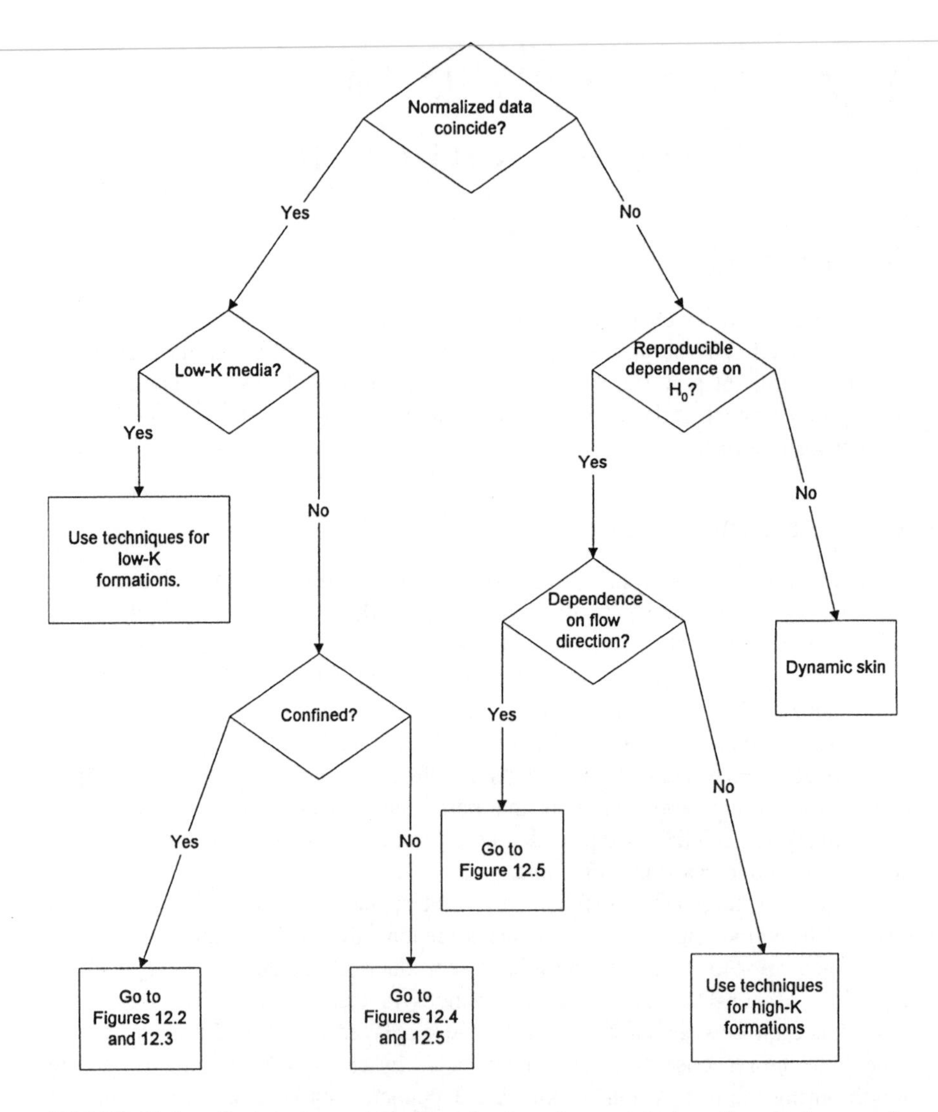

FIGURE 12.1 — Design tree to aid in the selection of a general class of methods to use for the analysis of response data from a slug test.

reflected in the size and type of the font used to designate that parameter. Three gradations are used in this and subsequent figures, which, in order from lowest to highest quality, are as follows: standard text font (e.g., S_s estimate of Figure 12.2), large font (e.g., K_r estimate obtained with the Peres et al. method in Figure 12.2), and large font with bold type (e.g., K_r estimates along the bottom of Figure 12.2). It is important to emphasize that these are only gradations in the relative quality of the parameter estimates, as all of the analysis methods are subject to error introduced by uncertainty regarding the effective screen length and other well-construction parameters.

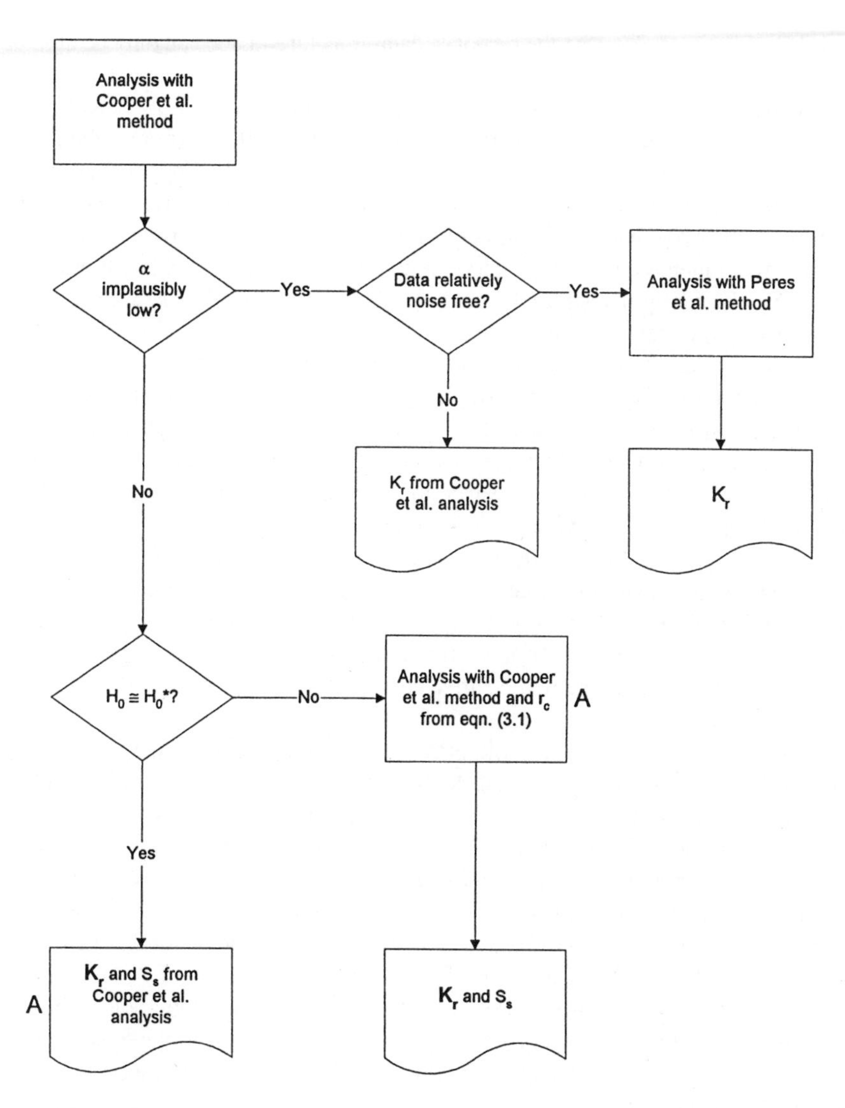

FIGURE 12.2 Flow chart to aid in the selection of analysis methodology for a slug test performed in a fully penetrating well in a confined formation.

Confined Formations — Fully Penetrating Wells

Figure 12.2 presents a flow chart for the selection of analysis methodology for slug tests performed in fully penetrating wells in confined formations. In all cases, the response data should initially be analyzed with the Cooper et al. method. If a very close match (e.g., Figure 5.4b and Figure 10.8 [well Ln-2]) can be obtained between the response data and a Cooper et al. type curve with a physically plausible α, then one can assume that skin effects will not have a major impact on the analysis. If a close match cannot be obtained with a physically plausible α, then one should assume that skin effects are affecting the response data. If a close match can only be obtained with an α estimate that is lower than what would be physically plausible, an attempt should be made to apply the approximate deconvolution method of Peres et al. If that approach is not viable, owing to limitations in the response data (e.g., Figure 5.20), the hydraulic conductivity value obtained with the Cooper et al. method should be considered as the best available estimate. That estimate, however, may only be a rather rough approximation of the conductivity of the formation as a result of the nonuniqueness of the type curve match. If the α estimate is larger than what would be physically plausible, the analysis should proceed as if skin effects were not affecting the response data.

If skin effects do not appear to be having a significant effect on the response data, the next step is to compare the measured (H_0) and expected (H_0^*) values for the initial displacement. If these two quantities are in close agreement, then a relatively good estimate of the conductivity of the formation should be obtained with the Cooper et al. method. However, the nonuniqueness of the type curve match will result in a specific storage estimate of lower quality.

If H_0 is considerably less than H_0^*, the response data should be analyzed with the Cooper et al. model using an effective casing radius calculated from the mass balance of Equation (3.1). This procedure should result in a relatively good estimate of the hydraulic conductivity of the formation. Once again, however, the quality of the specific storage estimate will be affected by the nonuniqueness of the type curve match.

The Cooper et al. method is the primary technique recommended here for the analysis of response data from slug tests performed in fully penetrating wells in confined formations because of its ability to detect the existence of a low-conductivity well skin, which is reflected in an implausibly low α estimate. In the absence of skin effects, the Hvorslev method will also produce very reasonable estimates of the hydraulic conductivity of the formation. Thus, this technique could readily be substituted for the Cooper et al. method at the points designated by the letter "A" on Figure 12.2. Unfortunately, however, the presence of a well skin will be difficult to recognize with the Hvorslev method. Given the magnitude of the error that can potentially be introduced through a failure to recognize the existence of a low-conductivity skin, the Cooper et al. method should always be used as at least a preliminary screening tool.

Confined Formations — Partially Penetrating Wells

Figure 12.3 presents a flow chart for the selection of analysis methodology for slug tests performed in partially penetrating wells in confined formations. As in the fully penetrating case, the response data should initially be analyzed with the Cooper et al. method. If a very close match (e.g., Figure 5.4b and Figure 10.8 [well Ln-2]) can be obtained between the response data and a Cooper et al. type curve with a physically plausible α, then one can assume that the vertical component of hydraulic conductivity is much less than the radial component, i.e., flow is constrained to the vertical interval of the formation intersecting the well screen. In that case, the analysis process should follow the steps outlined in Figure 12.2.

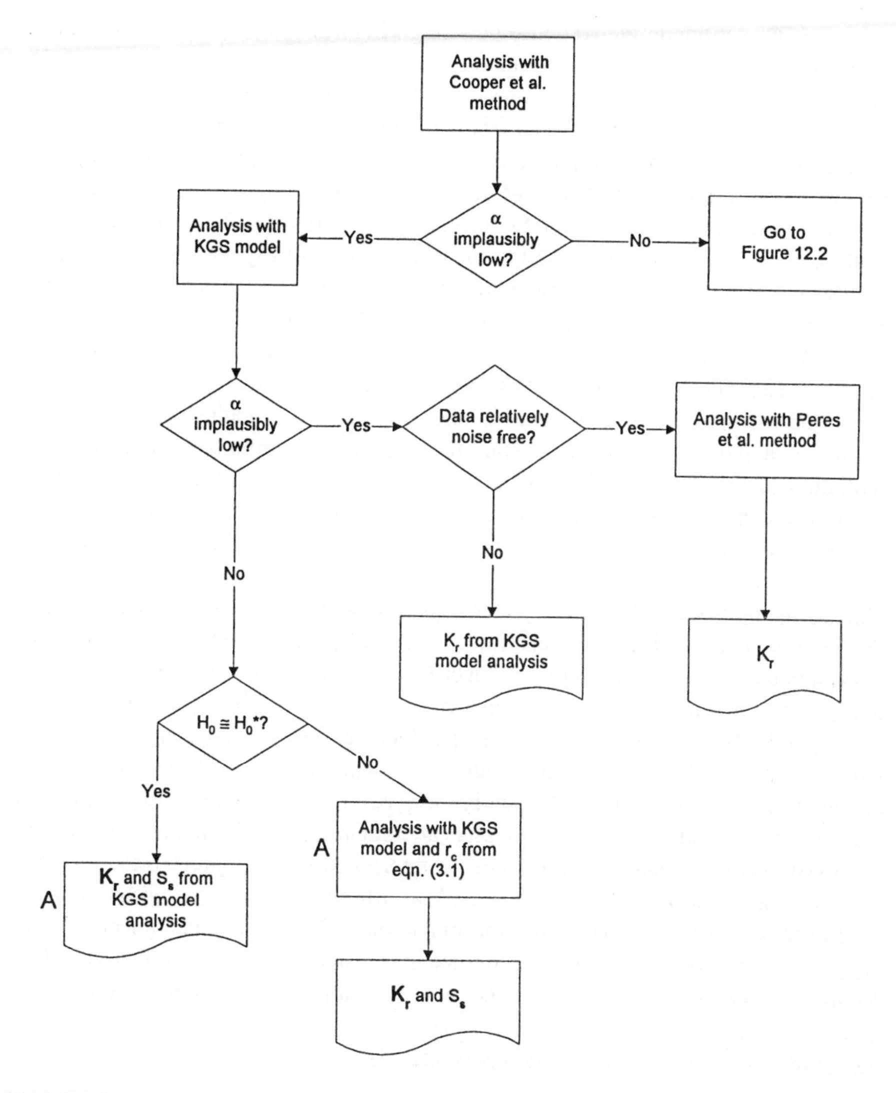

FIGURE 12.3 Flow chart to aid in the selection of analysis methodology for a slug test performed in a partially penetrating well in a confined formation.

If a very close match cannot be obtained using the Cooper et al. method with a physically plausible α (e.g., Figure 5.14), the response data should be analyzed with the isotropic form of the KGS model. If a very close match can be obtained between the response data and a KGS model type curve with a physically plausible α, then skin effects should not have a major impact on the analysis. If a close match cannot be obtained using a physically plausible α, then skin effects are very likely affecting the response data. If a close match can only be obtained with an α estimate that is lower than what would be physically plausible, an attempt should be made to apply the approximate deconvolution method of Peres et al. If that approach is not viable, owing to limitations in the response data (e.g., Figure 5.20), the hydraulic conductivity value obtained with the KGS model should be considered as the best available estimate. That estimate, however, may only be a rather rough approximation of the conductivity of the formation as a result of the nonuniqueness of the type curve match. If the α estimate is larger than what would be physically plausible, the analysis should proceed as if skin effects were not affecting the response data.

If skin effects do not appear to be having a significant effect on the response data, the next step is to compare the measured (H_0) and expected (H_0^*) values for the initial displacement. If these two quantities are in close agreement, then a relatively good estimate of the conductivity of the formation should be obtained with the KGS model. However, the nonuniqueness of the type curve match will result in a specific storage estimate of lower quality.

If H_0 is considerably less than H_0^*, the response data should be analyzed with the KGS model and an effective casing radius calculated from the mass balance of Equation (3.1). This procedure should result in a very reasonable estimate of the hydraulic conductivity of the formation. Once again, however, the quality of the specific storage estimate will be affected by the nonuniqueness of the type curve match.

The KGS model is the primary technique recommended here for the analysis of response data from slug tests performed in partially penetrating wells in confined formations because of its ability to detect the existence of a low-conductivity well skin, which is reflected in an implausibly low α estimate. In the absence of skin effects, the Hvorslev method will also produce relatively good conductivity estimates. Thus, this technique could readily be substituted for the KGS model at the points designated by the letter "A" on Figure 12.3. Although the confined variant of the Dagan method is an equally viable alternative to the Hvorslev method, the Hvorslev method is easier to use since no interpolation is required. Unfortunately, however, the presence of a well skin will be difficult to recognize with either the Hvorslev or Dagan method. Given the magnitude of the error that can potentially be introduced through a failure to recognize the existence of a low-conductivity skin, the KGS model should always be used as at least a preliminary screening tool.

Unconfined Formations — Wells Screened below the Water Table

Figure 12.4 presents a flow chart for the selection of analysis methodology for slug tests performed in wells screened below the water table in unconfined formations. In all cases, the response data should initially be analyzed with the Cooper et al. method. If a very close match (e.g., Figure 5.4b and Figure 10.8 [well Ln-2]) can be obtained between the response data and a Cooper et al. type curve with a physically plausible α, then one can assume that the vertical component of hydraulic conductivity is much less than the radial component, i.e., flow is constrained to the vertical interval of the formation intersecting the well screen. In that case, the analysis process should follow the steps outlined in Figure 12.2.

If a very close match cannot be obtained using the Cooper et al. method with a physically plausible α (e.g., Figure 5.14), the response data should be analyzed with

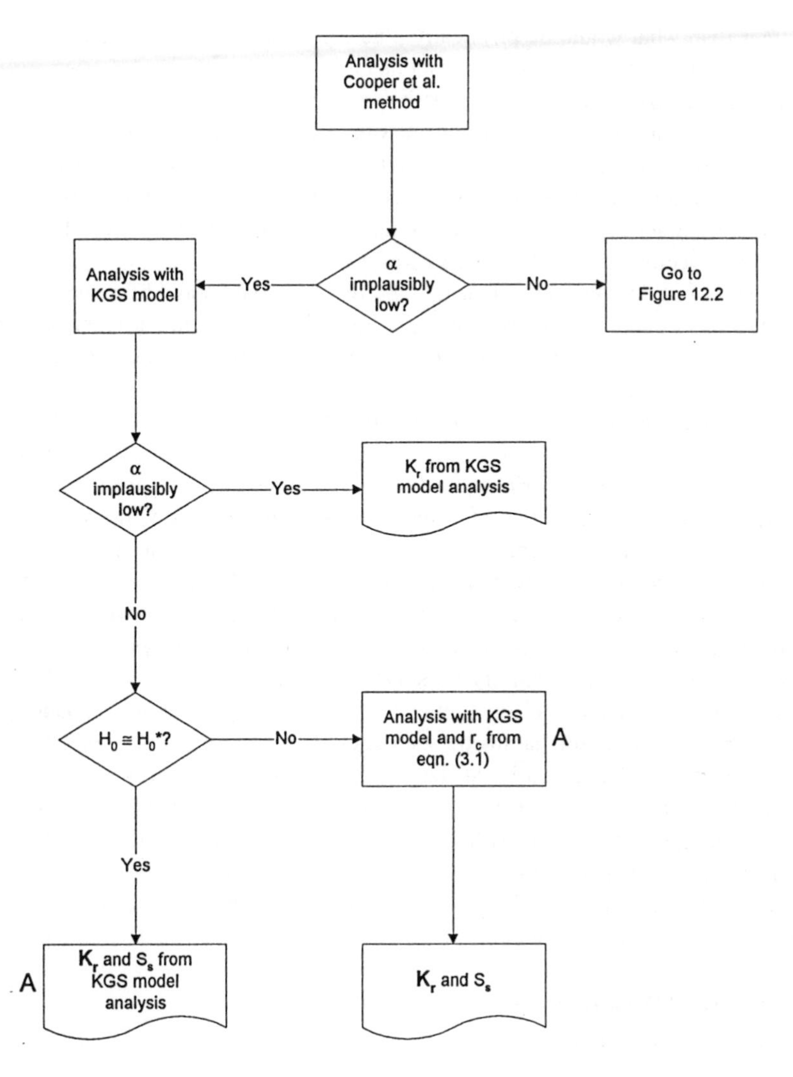

FIGURE 12.4 Flow chart to aid in the selection of analysis methodology for a slug test performed in a well screened below the water table in an unconfined formation.

the isotropic form of the KGS model. If a very close match can be obtained between the response data and a KGS model type curve with a physically plausible α, then one can assume that skin effects will not have a major impact on the analysis. If a close match cannot be obtained with a physically plausible α, then one should assume that skin effects are affecting the response data. If a close match can only be obtained with an α estimate that is lower than what would be physically plausible, the analyst has few options. In this case, the hydraulic conductivity value obtained with the KGS model should be considered as the best available estimate. That estimate, however, may only be a rather rough approximation of the conductivity of the formation as a result of the nonuniqueness of the type curve match. If the α estimate is larger than what would be physically plausible, the analysis should proceed as if skin effects were not affecting the response data.

If skin effects do not appear to be having a significant impact on the response data, the next step is to compare the measured (H_0) and expected (H_0^*) values for the initial displacement. If these two quantities are in close agreement, then a relatively good estimate of the conductivity of the formation should be obtainable with the KGS model. However, the nonuniqueness of the type curve match will result in a specific storage estimate of lower quality.

If H_0 is considerably less than H_0^*, the response data should be analyzed with the KGS model and an effective casing radius calculated from the mass balance of Equation (3.1). This procedure should result in a relatively good estimate of the hydraulic conductivity of the formation. Once again, however, the quality of the specific storage estimate will be affected by the nonuniqueness of the type curve match.

The KGS model is the primary technique recommended here for the analysis of response data from slug tests performed in wells screened below the water table in unconfined formations because of its ability to detect the existence of a low-conductivity well skin, which is reflected in an implausibly low α estimate. In the absence of skin effects, the Bouwer and Rice or Dagan method will also produce relatively good estimates of the hydraulic conductivity of the formation. Thus, either of these techniques could readily be substituted for the KGS model at the points designated by the letter "A" on Figure 12.4. Although the method of Dagan will produce somewhat better parameter estimates than that of Bouwer and Rice, the difference in the relative quality of the estimates produced by these two methods is of little practical significance. Unfortunately, however, the presence of a well skin will be difficult to recognize with either of these techniques. Given the magnitude of the error that can potentially be introduced through a failure to recognize the existence of a low-conductivity skin, the KGS model should always be employed as at least a preliminary screening tool.

UNCONFINED FORMATIONS — WELLS SCREENED ACROSS THE WATER TABLE

Figure 12.5 presents a flow chart for the selection of analysis methodology for slug tests performed in wells screened across the water table. In this configuration, the primary determinant of analysis methodology is the presence or absence of a reproducible dependence on H_0. If the response data do not exhibit a reproducible dependence on H_0, the analysis method will be selected in a manner similar to that outlined in Figure 12.4. The one exception to this is if drainage of the filter pack produces a response similar to that of Figure 6.6. In that case, the double straight line approach should be used to analyze the response data with the Bouwer and Rice method and an effective casing radius estimated from Equation (6.10) or (6.11b).

If the response data do exhibit a reproducible dependence on H_0, the Dagan method should be used in the analysis. The details of the application of the Dagan method will depend on the results of a comparison of the measured (H_0) and expected

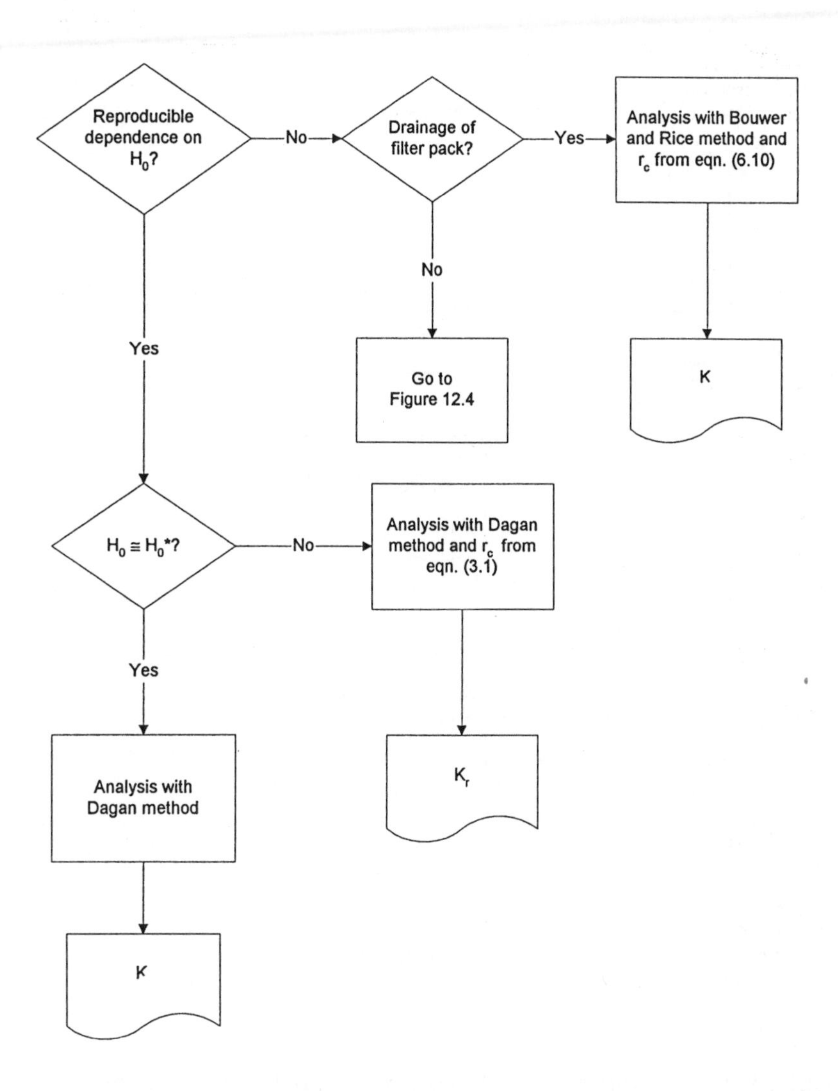

FIGURE 12.5 Flow chart to aid in the selection of analysis methodology for a slug test performed in a well screened across the water table.

(H_0^*) values for the initial displacement. If drainage of the filter pack is also a significant mechanism, Equation (6.10) or (6.11b) should be employed to calculate the effective casing radius.

Unfortunately, the existence of a low-permeability well skin is difficult to recognize with either the Bouwer and Rice or Dagan method; so, the estimates obtained from the analysis of tests in this configuration must be considered of lower quality than those obtained from configurations in which preanalysis screening can be used to recognize the presence of a low-permeability skin.

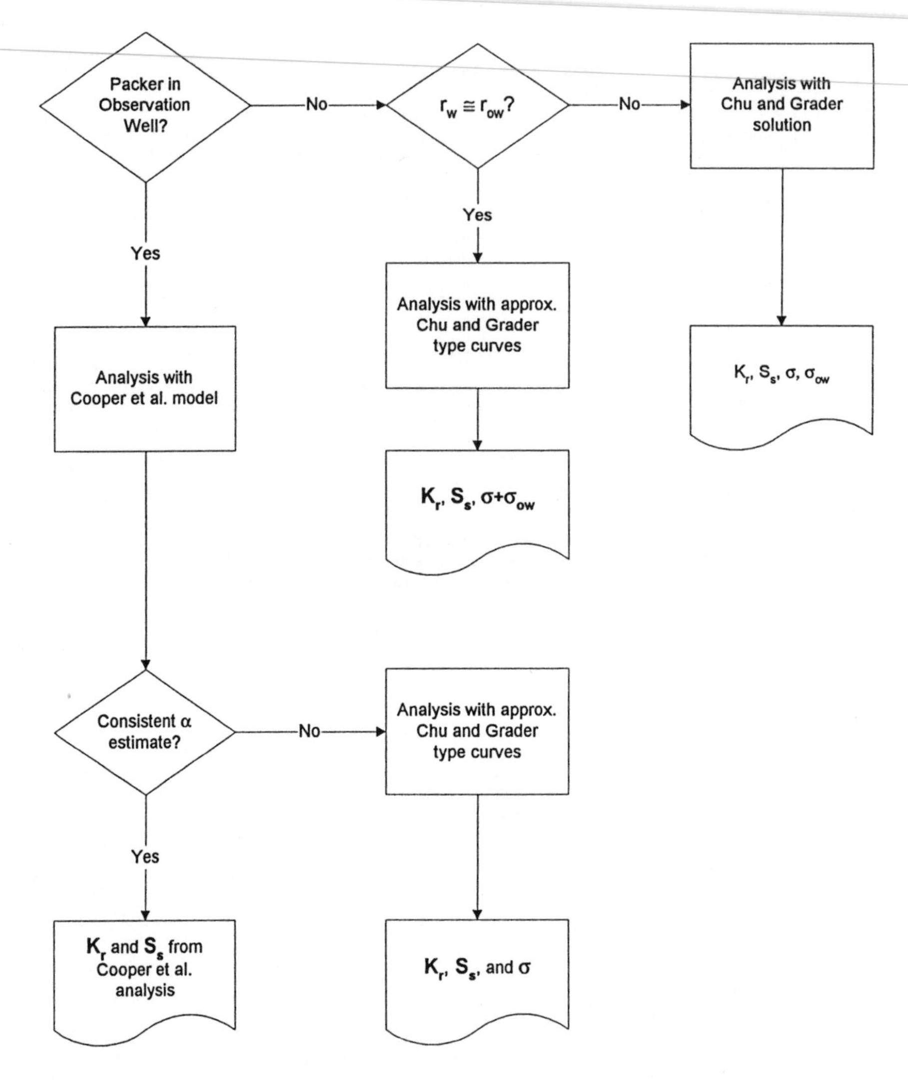

FIGURE 12.6 Flow chart to aid in the selection of analysis methodology for a multiwell slug test performed in a fully penetrating well in a confined formation.

Low Conductivity Formations

The process of selection of a technique for the analysis of slug tests performed in wells in low-conductivity formations is quite similar to the procedures outlined in Figures 12.2 to 12.5. The only major difference arises from the very small effective casing radius used in shut-in tests, which results in a very large α value. The normalized head vs. logarithm of $\alpha\beta$ type curves (Figure 7.6) may be the most appropriate approach for analysis of response data at very large α. In all cases, however, the response data should initially be analyzed using the procedures outlined in Figures 12.2-12.5. If an analysis with the Cooper et al. or KGS model yields an

α value greater than about 0.1, the $\alpha\beta$ type curves should be utilized for the analysis. Note that Figures 12.2 to 12.5 may not be applicable for tests in formations of extremely low conductivity as a result of the impact of the nonlinear mechanisms discussed in Chapter 7.

High Conductivity Formations

The Springer and Gelhar model or the linear variant of the McElwee et al. model should be the primary methods for the analysis of slug tests performed in highly permeable formations. If there is a possibility that noninstantaneous slug introduction or nonlinear mechanisms are affecting the response data, the translation method should be applied to the data prior to the analysis as described in Chapter 8. The effect of nonlinear mechanisms on estimates obtained with the translation method can be assessed by comparing results from tests in which H_0 varied by a factor of two or more. If the impact of nonlinear mechanisms still appears significant, the nonlinear variant of the McElwee et al. model should be used in the analysis.

Multiwell Tests

Figure 12.6 presents a flow chart for the selection of analysis methodology for multiwell slug tests performed in fully penetrating wells in confined formations. In this case, the presence or absence of a packer in the observation well is one of the primary determinants of analysis methodology. If a packer is used in the observation well, the response data from both the test and observation wells should initially be analyzed with the Cooper et al. method. If a very close match is obtained between these data and type curves for a single α (e.g., Figure 10.8), then one can assume that skin effects will not have a major impact on the analysis. If a close match cannot be obtained, then the response data at the observation well should be analyzed with the approximate type curve method of Chu and Grader. In either case, relatively good estimates of the hydraulic properties of the formation can be obtained.

If a packer is not used in the observation well, a comparison of the effective screen radii for the test and observation wells will be the primary determinant of analysis methodology. If the effective screen radii are approximately equal, wellbore storage effects at the two wells should be similar and the approximate type curve method of Chu and Grader can be used to analyze the response data. If the two radii are not similar, the general solution of Chu and Grader must be used in the analysis. In that case, however, parameter nonuniqueness can introduce a considerable degree of uncertainty into the estimates of the hydraulic properties of the formation.

For multiwell tests initiated in partially penetrating wells, a packer should be placed in the observation well and the multiwell variant of the KGS model used for the analysis. The presence of a skin at the test well can be assessed in a manner similar to that used with the Cooper et al. method.

13 Final Comments

Is the slug test a reasonable means for obtaining information about the transmissive and storage properties of a formation? Hopefully, the reader will now feel comfortable in joining me in replying to this question with a qualified "yes". One of the objectives of this book was to demonstrate that a program of slug tests can provide valuable information about the hydraulic properties of a formation. Although not all readers might find the presentation quite so convincing, I am confident that there will be near-unanimous agreement on at least one point. That is, the value of the information obtained from a series of slug tests will largely be determined by the details of the procedures followed in the design, performance, and analysis phases of the test program. Clearly, if proper attention is not paid to the details of these activities, the information obtained from a slug test may bear little resemblance to reality.

Three themes were defined in Chapter 1 and then reiterated throughout the book. I believe that it is worthwhile here, in the final chapter, to briefly summarize the major findings concerning these themes, with an emphasis on the quality of the parameter estimates that can be expected from a slug test.

1. Importance of well development — if development activities are moderately effective at a well, I believe that it is reasonable to expect that the hydraulic conductivity estimate from a slug test will be within a factor of two of the conductivity of the formation in the vicinity of the test well. However, if there has been little to no development, the conductivity estimate can easily be in error by an order of magnitude or much more. Well construction procedures will often largely determine the effectiveness of development activities;
2. Importance of test design — proper test design is critical for assessing which class of methods should be used to analyze the response data. Inattention to test design, whether it result in, for example, a linear model being used for tests affected by nonlinear mechanisms or an inappropriate estimate for the effective casing radius, can easily lead to a conductivity estimate that is in error by a factor of three to four;
3. Importance of appropriate analysis procedures — the Hvorslev and Bouwer and Rice methods can be the source of significant errors, easily an order of magnitude or more, when they are used to analyze tests performed in the presence of a low-permeability well skin. Estimates obtained with these commonly used methods must therefore be viewed with healthy skepticism unless there is some evidence, such as the results of preliminary screening analyses, that skin effects can be neglected.

I trust that the ample discussion and field examples presented in the text have convinced the reader of the importance of these themes.

As should be evident from the preceding summary, the single-most important element of a program of slug tests is well development. Unfortunately, regardless of the efforts and motivations of the individuals involved in the field investigation, the effects of incomplete well development may be difficult to avoid. Thus, I strongly recommend that the hydraulic conductivity estimate obtained from a program of slug tests always be viewed as a lower bound on the conductivity of the formation in the vicinity of the well. This bounding value will often be a very reasonable estimate of the hydraulic conductivity of the formation when appropriate development procedures are utilized.

"But aren't there any better alternatives?" is certainly a logical question to ask given some of the limitations of slug-test methodology. In many situations, a pumping test is clearly a superior alternative to a slug test, as it can often provide parameter estimates that are not affected by conditions in the near-well zone of disturbance. Unfortunately, however, various boundary effects, ranging from wellbore storage to leakage from adjoining units, can greatly complicate the analysis of pumping-induced drawdown. When this is considered along with the logistical disadvantages of the technique, which are especially pronounced at sites of suspected groundwater contamination or in low-conductivity formations, it becomes evident that the pumping test has some fairly significant limitations of its own. A wide variety of other methods, such as grain-size analyses, correlations with geophysical logs, etc., can also be used to obtain estimates of the hydraulic conductivity of the formation. However, I doubt that many of these approaches can provide estimates that are really much better than what an experienced hydrogeologist can produce with a lithologic description, i.e., an estimate within one to two orders of magnitude.

A slug test may, therefore, be one of the most appropriate approaches for estimation of hydraulic conductivity at many sites. A properly designed, performed, and analyzed series of slug tests can yield estimates that are quite reasonable representations of the bulk hydraulic properties of the formation. It is my hope that this book has provided the reader with the information necessary to better realize the full potential of this class of field methods. Good luck!

Appendix A: A Brief Discussion of Analysis Software

There are presently a large number of commercially available software packages for the analysis of data from various types of hydraulic tests. Most of these packages will have some capabilities to analyze response data from slug tests. For the most part, these capabilities are limited to analyses using the Cooper et al., Hvorslev, and Bouwer and Rice methods. The Hvorslev and Bouwer and Rice methods are usually performed with the fitting procedure outlined in Figure 5.8A. The KGS model is currently not available in most packages. A copy of the FORTRAN code implementing the KGS model is available from the Publication and Sales Office of the Kansas Geological Survey (1930 Constant Ave., Campus West, University of Kansas, Lawrence, KS 66047, (785) 864-3965) at a nominal charge. Computer Program Series 95-1 (Liu and Butler, 1995) should be requested. The documented FORTRAN code, an executable file, and a series of examples are given in that report. No automated analysis capabilities are included, but the program can be used to generate type curves for graphical curve matching. Note that this program can also be used to generate the Cooper et al. type curves, the $\alpha\beta$ type curves of Figure 7.6, and the $\alpha_{lf}\beta_{lf}$ type curve of Figure 11.2. In all cases, type curves can be generated for responses at both the test well and any observation wells. A packer must have been used in the observation well for these type curves to be appropriate.

Few commercially available packages currently offer options for the analysis of response data from slug tests in high-conductivity formations. However, analyses with the Springer and Gelhar method and the linearized variant of the McElwee et al. method can be readily performed with more general mathematical analysis software. For example, the type curves of Figure 8.6 and the fit of Figure 8.8 were produced with the Mathcad software package (MathSoft, 1995).

Most analysis packages used by hydrogeologists do not offer options for the calculation of head derivatives or equivalent drawdown data. However, these calculations can be readily performed with the DERIV program of Spane and Wurstner (1993). This program can be obtained by writing Frank Spane at Pacific Northwest Laboratory, P.O. Box 999, Richland, WA 99352.

Appendix B: List of Notation

- ***A*,*B*,*C*** empirical coefficients used in the Bouwer and Rice method, [dimensionless];
- A dimensionless parameter of nonlinear term in equation for flow in high-conductivity formations or cross-sectional area term used in linear-flow model for fractured formations [L];
- B formation thickness, [L];
- b screen length or intercept of transducer conversion equation, [L];
- b_f fracture aperture, [L];
- C damping parameter for slug tests in high-conductivity formations, [T^{-1}];
- C_{BK} shape factor of Boast and Kirkham, [dimensionless];
- C_d dimensionless damping parameter for slug tests in high-conductivity formations;
- C_{ef} effective compressibility term, [LT^2/M];
- d z position of end of screen closest to an impermeable boundary in confined formation or to water table in unconfined formation, [L];
- F viscous loss parameter, [L/T];
- g gravitational acceleration, [L/T^2];
- H deviation of hydraulic head in test well from static conditions, [L];
- H_0 measured value for the initial displacement determined immediately after test initiation, [L];
- H_0^* expected value for the initial displacement determined from volumetric considerations or pre-test measurements, [L];
- H_0^+ apparent value for the initial displacement determined from the y-intercept of the best-fit straight line, [L];
- H_{md} modified normalized head used in Chu and Grader method, [dimensionless];
- H_{ow}, h_{ow} deviation of hydraulic head in the observation well from static conditions, [L];
- h deviation of hydraulic head in the formation from static conditions, [L];
- hp_i pressure head produced by the overlying column of water, [L];
- hp_{st} pressure head at static conditions, [L];
- h_s normalized head at shut-in for drillstem test, [dimensionless];
- K_r radial component of the hydraulic conductivity of the formation, [L/T];
- K_{sk} hydraulic conductivity of skin, [L/T];
- K_z vertical component of the hydraulic conductivity of the formation, [L/T];
- L nominal length of the water column in well, [L];
- L_e effective length of the water column in well, [L];
- m slope of transducer conversion equation ([L/I] or [L/V]) or the slope of the normalized head vs. log(t/t′) plot used for drillstem tests, [dimensionless];
- m_q the slope of the normalized head over q_{av} vs. log(t/t′) plot used for drillstem tests, [T/L^2];

n drainable porosity of the filter pack, [dimensionless];
P dimensionless flow parameter used in Dagan method;
Q pumping rate, $[L^3/T]$;
Q_D dimensionless flow rate term of Karasaki (1990) used for drillstem tests;
q_{av} the average flow rate during the flow period of a drillstem test, $[L^2/T]$;
q_D dimensionless flow rate from the formation used in the approximate deconvolution method;
R_{Dmax} the maximum radial extent of a slug test divided by r_w of the test well, [dimensionless];
R_e effective radius parameter, [L];
r radial direction, [L];
r_c effective radius of well casing, [L];
r_c' effective casing radius for a shut-in slug test, [L];
r_D r_{obs}/r_w, [dimensionless];
r_{nc} nominal casing radius, [L];
r_{obs} radial distance from test well to observation well, [L];
r_{ow} effective screen radius of observation well, [L];
r_{owc} effective casing radius of observation well, [L];
r_{sk} outer radius of well skin, [L];
r_w effective radius of well screen, [L];
r_w^* $r_w(K_z/K_r)^{1/2}$;
S_s specific storage of the formation, [1/L];
$S_s e^{-2\sigma}$ apparent specific storage, [1/L];
s pumping-induced drawdown, [L];
$\tilde{s}$ equivalent drawdown for the case of negligible wellbore storage, [T];
$\Delta\tilde{s}$ change in $\tilde{s}$ over a log cycle in time, [T];
s_D $(4\pi K_r Bs/Q)$, dimensionless drawdown;
$\tilde{s}_D$ dimensionless drawdown for case of negligible wellbore storage;
T_0 basic time lag, time at which a normalized head of 0.368 is obtained, [T];
t total time since the start of the test, [T];
t' time since shut-in for a drillstem test, [T];
t_0 time of test initiation, [T];
$t_{1.0}$ real time corresponding to $\beta = 1.0$, [T];
t_d $(g/L_e)^{0.5}t$, [dimensionless];
t_n time of the nth peak or trough in the test data, [T];
t_s time of shut-in for a drillstem test, [T];
V_w volume of water below the packer, $[L^3]$;
W_n the w value at the nth peak or trough in the test data, [L];
w deviation of water level from static level in the test well, [L];
w_d w/H_0, [dimensionless];
x_i current or voltage measurement from transducer, [I] or [V];
x_{st} current or voltage measurement from transducer corresponding to static conditions, [I] or [V];
x_0 current or voltage measurement from transducer corresponding to pressure head equivalent to H_0, [I] or [V];
z vertical direction, [L];

α	$(r_w^2 S_s B)/r_c^2$, dimensionless storage parameter, b replaces B in partially penetrating models;
α_{cal}	α value calculated via curve matching;
α^*	$\alpha e^{-2\sigma}$, [dimensionless];
α_{ow}	$(r_{ow}^2 S_s B)/r_{owc}^2$, dimensionless storage parameter for the observation well;
α_1	dimensionless storage parameter of the open-hole phase of a drillstem test;
α_2	dimensionless storage parameter of the shut-in phase of a drillstem test;
α_{lf}	$(Ar_w^2 S_s)/r_c^2$, dimensionless storage parameter for the linear-flow model;
β	$K_r Bt/r_c^2$, dimensionless time parameter, b replaces B in partially penetrating models;
β_{lf}	$(AKt)/(r_c^2)$, dimensionless time parameter for the linear-flow model;
γ	correlating parameter used in Chu and Grader method, [dimensionless];
λ	interporosity flow parameter, characterizes ease of fluid movement between matrix and fractures, [dimensionless];
ν	kinematic viscosity, $[L^2/T]$;
ρ_w	density of water, $[M/L^3]$;
σ	$((K_r/K_{sk}) - 1)\ln(r_{sk}/r_w)$, skin factor at the test well, [dimensionless];
σ_{ow}	skin factor at the observation well, [dimensionless];
τ	$4\beta/\alpha$, dimensionless time definition used in the approximate deconvolution method of Peres et al.;
τ_{md}	modified dimensionless time used in Chu and Grader method, [dimensionless];
ψ	$\frac{\sqrt{K_z/K_r}}{b/r_w}$, dimensionless;
ω	frequency parameter for slug tests in high-conductivity formations ($[T^{-1}]$) or dimensionless double-porosity storage parameter;
$\Box(z)$	boxcar function $= 0,\ z < d,\ z > b + d,$ $= 1$, elsewhere.

References

Aller, L., Bennett, T. W., Hackett, G., Petty, R. J., Lehr, J. H., Sedoris, H., Nielsen, D. M., and Denne, J. E., *Handbook of Suggested Practices for the Design and Installation of Ground-Water Monitoring Wells*, National Water Well Assoc., Dublin, 1989.

Amoozegar, A. and Warrick, A. W., Hydraulic conductivity of saturated soils: Field methods, in *Methods of Soil Analysis, Part 1. Physical and Mineralogical Methods*, Agronomy Monograph Series 9, Klute, A., Ed., American Soc. of Agronomy, Madison, 1986, 735.

Aron, G., and Scott, V. H., Simplified solution for decreasing flow in wells, *ASCE J. Hydr. Engrg.*, 91(HY5), 1, 1965.

ASTM, Standard guide for development of groundwater monitoring wells in granular aquifers (D 5521-94), in *1996 Annual Book of ASTM Standards*, 4.09, American Soc. for Testing and Materials, Philadelphia, 1996, 344.

Barker, J. A., A generalized radial flow model for hydraulic tests in fractured rock, *Water Resour. Res.*, 24(10), 1796, 1988.

Barker, J. A., and Black, J. H., Slug tests in fissured aquifers, *Water Resour. Res.*, 19(6), 1558, 1983.

Beauheim, R. L., Interpretation of single-well hydraulic tests conducted at and near the Waste Isolation Pilot Plant (WIPP) Site, 1983-1987, *Sandia National Laboratory Rept. SAND87-0039*, Albuquerque, 1987.

Beauheim, R. L., Practical considerations in well testing of low-permeability media (abstract), *EOS*, 75(16) 151, 1994.

Bird, R. B., Stewart, W. E., and Lightfoot, E. N., *Transport Phenomena*, John Wiley & Sons, New York, 1960.

Black, J. H., The use of slug tests in groundwater investigations, *Water Services*, March, 174, 1978.

Black, J. H., The interpretation of slug tests in fissured rocks, *Quart. J. Eng. Geol.*, 18, 161, 1985.

Black, J. H., and Kipp, K. L., Jr., Observation well response time and its effect upon aquifer test results, *J. Hydrol.*, 34, 297, 1977a.

Black, J. H., and Kipp, K. L., Jr., The significance and prediction of observation well response delay in semiconfined aquifer-test analysis, *Ground Water*, 15(6), 1977b.

Boak, R. A., *Auger Hole, Piezometer, and Slug Tests: A Literature Review*, M.Sc. thesis, Civil Eng. Dept., Univ. of Newcastle upon Tyne, U.K., 1991.

Boast, C. W., and Kirkham, D., Auger hole seepage theory, *Soil Sci. Soc. Am. Proc.*, 35(3), 365, 1971.

Bourdet, D., Whittle, T. M., Douglas, A. A., and Pirard, Y. M., A new set of type curves simplifies well test analysis, *World Oil*, 196(6), 95, 1983.

Bouwer, H., The Bouwer and Rice slug test — an update, *Ground Water*, 27(3), 304, 1989.

Bouwer, H. and Jackson, R. D., Determining soil properties, in *Drainage for Agriculture*, van Schilfgaarde, J., Ed., Agron. Monogr., 17, 1974, 611.

Bouwer, H., and Rice, R. C., A slug test for determining hydraulic conductivity of unconfined aquifers with completely or partially penetrating wells, *Water Resour. Res.*, 12(3), 423, 1976.

Bredehoeft, J. D., The drill-stem test: The petroleum industry's deep-well pumping test, *Ground Water*, 3(1), 31, 1965.

Bredehoeft, J. D. and Papadopulos, I. S., A method for determining the hydraulic properties of tight formations, *Water Resour. Res.*, 16(1), 233, 1980.

Bredehoeft, J. D., Cooper, H. H., Jr., and Papadopulos, I. S., Inertial and storage effects in well-aquifer systems: An analog investigation, *Water Resour. Res.*, 2(4), 697, 1966.

Brother, M. R. and Christians, G. L., *In situ* slug test analysis: A comparison of three popular methods for unconfined aquifers, Proc. of the 7th National Outdoor Action Conf., *NGWA Ground Water Manage.*, 15, 597, 1993.

Butler, J. J., Jr., Pumping tests in nonuniform aquifers: The radially symmetric case, *J. Hydrol.*, 101, 15, 1988.

Butler, J. J., Jr., The role of pumping tests in site characterization: Some theoretical considerations, *Ground Water*, 28(3), 394, 1990.

Butler, J. J., Jr., Slug tests in site characterization: Some practical considerations, *Environ. Geosci.*, 3(3), 154, 1996.

Butler, J. J., Jr., Theoretical analysis of impact of incomplete recovery on slug tests, *Kans. Geol. Surv. Open-File Rep. 97-59*, 1997.

Butler, J. J., Jr. and Healey, J. M., Analysis of 1994-95 hydraulic tests at Trego County monitoring site, *Kans. Geol. Surv. Open-File Rep. 95-66*, 1995.

Butler, J. J., Jr. and Healey, J. M., Relationship between pumping-test and slug-test parameters: Scale effect or artifact? *Ground Water,* in press, 1998.

Butler, J. J., Jr. and Hyder, Z., An assessment of the Nguyen and Pinder method for slug test analysis, *Ground Water Monitoring Remediation*, 14(4), 124, 1994.

Butler, J. J., Jr. and Liu, W. Z., Analysis of 1991-1992 slug tests in the Dakota aquifer of central and western Kansas, *Kans. Geol. Surv. Open-File Rep. 93-1c*, 1997.

Butler, J. J., Jr. and McElwee, C. D., Well-testing methodologies for characterizing heterogeneities in alluvial-aquifer systems: Final technical report, *Kans. Geol. Surv. Open-File Rep. 95-75*, 1996.

Butler, J. J., Jr., Bohling, G. C., Hyder, Z., and McElwee, C. D., The use of slug tests to describe vertical variations in hydraulic conductivity, *J. Hydrol.*, 156, 136, 1994.

Butler, J. J., Jr., Liu, W. Z., and Young, D. P., Analysis of October 1993 slug tests in Stafford, Pratt, and Reno counties, south-central Kansas, *Kans. Geol. Surv. Open-File Rep. 93-52*, 1993

Butler, J. J., Jr., McElwee, C. D., and Liu, W. Z., Improving the quality of parameter estimates obtained from slug tests, *Ground Water*, 34(3), 480, 1996.

Campbell, M. D., Starrett, M. S., Fowler, J. D., and Klein, J. J., Slug tests and hydraulic conductivity, *Ground Water Manage.*, 2, 85, 1990.

Carslaw, H. S. and Jaeger, J. C., *Conduction of Heat in Solids*, Oxford Univ. Press, New York, 1959.

Chakrabarty, C. and Enachescu, C., Using the deconvolution approach for slug test analysis: Theory and application, *Ground Water*, in press, 1997.

Chirlin, G. R., A critique of the Hvorslev method for slug test analysis: The fully penetrating well, *Ground Water Monitor. Rev.*, 9(2), 130, 1989.

Chirlin, G. R., The slug test: the first four decades, *Ground Water Manage.*, 1, 365, 1990.

Chu, L. and Grader, A. S., Transient-pressure analysis for an interference slug test, *SPE Paper 23444*, Soc. of Pet. Engs., 1991.

Cole, K. D. and Zlotnik, V. A., Modification of Dagan's numerical method for slug and packer test interpretation, in *Computational Methods in Water Resources*, X, Peters, A., Wittum, G., Herrling, B., Meissner, U., Brebbia, C. A., Gray, W. G., and Pinder, G. F., Eds., Kluwer Academic, Dordrecht, 1994, 719.

Connell, L. D., The importance of pulse duration in pulse test analysis, *Water Resour. Res.*, 30(8), 2403, 1994.

Cooper, H. H., Bredehoeft, J. D., and Papadopulos, I. S., Response of a finite-diameter well to an instantaneous charge of water, *Water Resour. Res.*, 3(1), 263, 1967.

Cooper, H. H. and Jacob, C. E., A generalized graphical method for evaluating formation constants and summarizing well-field history, *Eos Trans.*, American Geophysical Union, 27(4), 526, 1946.

Correa, A. C. and Ramey, H. J., Jr., A method for pressure buildup analysis of drillstem tests, *SPE Paper 16802*, Soc. of Pet. Engs., 1987.

Dachler, R., *Grundwasserstromung* (Groundwater Flow), Springer-Verlag, Vienna, 1936.

Dagan, G., A note on packer, slug, and recovery tests in unconfined aquifers, *Water Resour. Res.*, 14(5), 929, 1978.

Dagan, G., *Flow and Transport in Porous Formations*, Springer-Verlag, New York, 1989.

Dahl, S., and Jones, J., Evaluation of slug test data under unconfined conditions with exposed screens, and low permeability filter pack, Proc. 7th Natl. Outdoor Action Conference, National Ground Water Association, Dublin, Ohio, pp. 609–623, 1993.

Dax, A., A note on the analysis of slug tests, *J. Hydrol.,* 91, 153, 1987.

Domenico, P. A. and Schwartz, F. W., *Physical and Chemical Hydrogeology*, John Wiley & Sons, New York, 1990.

Dougherty, D. E. and Babu, D. K., Flow to a partially penetrating well in a double-porosity reservoir, *Water Resour. Res.*, 20(8), 1116, 1984.

Driscoll, F. G., *Groundwater and Wells*, Johnson Division, St. Paul, 1986.

Faust, C. R. and Mercer, J. W., Evaluation of slug tests in wells containing a finite-thickness skin, *Water Resour. Res.*, 20(4), 504, 1984.

Ferris, J. G. and Knowles, D. B., The slug-injection test for estimating the coefficient of transmissibility of an aquifer, in *U.S. Geological Survey Water-Supply Paper 1536-I*, R. Bentall (compiler), 1963, 299.

Freeze, R. A. and Cherry, J. A., *Groundwater*, Prentice-Hall, Englewood Cliffs, NJ, 1979.

Gelhar, L. W., *Stochastic Subsurface Hydrology*, Prentice-Hall, Englewood Cliffs, NJ, 1993.

Gladfelter, R.E., Tracy, G. W., and Wilsey, L. E., Selecting wells which will respond to production-stimulation treatment, *Drilling Production Practices*, American Pet. Inst., 1955, 117.

Grader, A. S. and Ramey, H. J., Jr., Slug-test analysis in double-porosity reservoirs, *SPE Formation Evaluation*, 3(2), 329, 1988.

Guyonnet, D., Mishra, S., and McCord, J., Evaluating the volume of porous medium investigated during a slug test, *Ground Water*, 31(4), 627, 1993.

Hantush, M. S., 1964, Hydraulics of wells, in *Advances in Hydrosciences*, 1, Chow, V. T., Ed., Academic Press, New York, 1964, 281.

Harvey, C. F., *Interpreting Parameter Estimates Obtained from Slug Tests in Heterogeneous Aquifers*, M.S. thesis, Applied Earth Sci. Dept., Stanford University, Palo Alto, CA, 1992.

Hess, K. M., Wolf, S. H., and Celia, M. A., Large-scale natural gradient tracer test in sand and gravel, Cape Cod, Massachusetts, 3. Hydraulic conductivity variability and calculated macrodispersivities, *Water Resour. Res.*, 28(8), 2011, 1992.

Hinsby, K., Bjerg, P. L., Andersen, L. J., Skov, B., and Clausen, E. V., A mini slug test method for determination of a local hydraulic conductivity of an unconfined sandy aquifer, *J. Hydrol.,* 136, 87, 1992.

Horne, R. N., *Modern Well Test Analysis*, Petroway, Palo Alto, CA, 1995.

Horner, D. R., Pressure build-up in wells, in *Proc. Third World Petroleum Congress*, II, Brill, E. J., Ed., Leiden, The Hague, 1951, 503.

Hvorslev, M. J., Time lag and soil permeability in ground-water observations, *U.S. Army Corps of Engrs. Waterways Exper. Sta. Bull no. 36*, 1951.

Hyder, Z. and Butler, J. J., Jr., Slug tests in unconfined formations: An assessment of the Bouwer and Rice technique, *Ground Water*, 33(1), 16, 1995.

Hyder, Z., Butler, J. J., Jr., McElwee, C. D., and Liu, W. Z., Slug tests in partially penetrating wells, *Water Resour. Res.*, 30(11), 2945, 1994.

Johnson, C. R., Greenkorn, R. A., and Woods, E. G., Pulse-testing: A new method for describing reservoir flow properties between wells, *J. Pet. Tech.*, 18, 1599, 1966.

Kabala, Z. J., Pinder, G. F., and Milly, P. C. D., Analysis of well-aquifer response to a slug test, *Water Resour. Res.*, 21(9), 1433, 1985.

Karasaki, K., A systematized drillstem test, *Water Resour. Res.*, 26(12), 2913, 1990.

Karasaki, K., Prematurely terminated slug tests: A field application, in *Proc. Int. Symp. Ground Water*, Lennon, G. P., Ed., Am. Soc. of Civil Eng., 1991, 163.

Karasaki, K., Long, J. C. S., and Witherspoon, P. A., Analytical models of slug tests, *Water Resour. Res.*, 24(1), 115, 1988.

Kell, G. S., Volume properties of ordinary water, in *Handbook of Chemistry and Physics*, 56th ed., Weast, R. C., Ed., CRC Press, Boca Raton, FL, 1975.

Keller, C. K. and Van der Kamp, G., Slug tests with storage due to entrapped air, *Ground Water*, 30(1), 2, 1992.

Kill, D. L., Monitoring well development — why and how, in *Ground Water and Vadose Monitoring — ASTM STP 1053*, Nielsen, D. M., and Johnson, A. I., Eds., American Soc. for Testing and Materials, Philadelphia, 1990, 82.

Kipp, K. L., Jr., Type curve analysis of inertial effects in the response of a well to a slug test, *Water Resour. Res.*, 21(9), 1397, 1985.

Krauss, I., Determination of the transmissibility from the free water level oscillation in well-aquifer systems, *Proc. Fort Collins Third Int. Hydrology Symp.*, Morel-Seytoux, H.J., Salas, J. D., Sanders, T. G., and Smith, R. E., Eds., 1977, 268.

Kreyszig, E., *Advanced Engineering Mathematics*, John Wiley & Sons, New York, 1979.

Krosynski, U. I. and Dagan, G., Well pumping in unconfined aquifers: The influence of the unsaturated zone, *Water Resour. Res.*, 11(3), 479, 1975.

Kruseman, G. P. and de Ridder, N. A., *Analysis and Evaluation of Pumping Test Data — ILRI Publication 47*, The Netherlands, Int. Inst. for Land Reclamation and Improvement, 1990.

Leap, D. I., A simple pneumatic device and technique for performing rising water level slug tests, *Ground Water Monitoring Rev.*, 4(4), 141, 1984.

Levy, B. S., Pannell, L. J., and Dadoly, J. P., A pressure-packer system for conducting rising head tests in water table wells, *J. Hydrol.*, 148, 189, 1993.

Liu, W. Z. and Butler, J. J., Jr., The KGS model for slug tests in partially penetrating wells (version 3.0), *Kans. Geol. Surv. Comput. Ser. Rep. 95-1*, 1995.

Luthin, J. N. and Kirkham, D., A piezometer method for measuring permeability of soil *in situ* below a water table, *Soil Sci.*, 68, 349, 1949.

Mateen, K. and Ramey, H. J., Jr., Slug test data analysis in reservoirs with double porosity behavior, *SPE Paper 12779*, Soc. of Pet. Engrs., 1984.

MathSoft, Inc., *Mathcad PLUS 6.0 User's Guide*, 1995.

McElwee, C. D. and Butler, J. J., Jr., Slug testing in highly permeable aquifers, *Kans. Geol. Surv. Open-File Rep. 89-28*, 1989.

McElwee, C. D. and Butler, J. J., Jr., Experimental verification of a general model for slug tests, *Kans. Geol. Surv. Open-File Rep. 96-47*, 1996.

McElwee, C. D. and Zenner, M., Unified analysis of slug tests including nonlinearities, inertial effects, and turbulence, *Kans. Geol. Surv. Open-File Rep. 93-45*, 1993.

McElwee, C. D., Butler, J. J., Jr., and Bohling, G. C., Nonlinear analysis of slug tests in highly permeable aquifers using a Hvorslev-type approach, *Kans. Geol. Surv. Open-File Rep. 92-39*, 1992.

McElwee, C. D., Bohling, G. C., and Butler, J. J., Jr., Sensitivity analysis of slug tests, I. The slugged well, *J. Hydrol.,* 164, 53, 1995a.

McElwee, C. D., Butler, J. J., Jr., Liu, W. Z., and Bohling, G. C., Sensitivity analysis of slug tests, II. Observation wells, *J. Hydrol.,* 164, 69, 1995b.

McGuire, V. and Zlotnik, V. A., Characterizing vertical distribution of horizontal hydraulic conductivity in an unconfined sand and gravel aquifer using double packer slug tests (abstract), *Ground Water,* 33(5), 850, 1995.

McLane, G. A., Harrity, D. A., and Thomsen, K. O., A pneumatic method for conducting rising and falling head tests in highly permeable aquifers, Proc. 1990 NWWA Outdoor Action Conf., National Water Well Assoc., 1990.

Mishra, S., Deconvolution Analysis of Slug Test Data from Composite Systems, unpublished, 1991.

Mishra, S., On Estimating Storativity from Slug-Test Data, unpublished, 1997.

Moench, A. F., Double-porosity models for a fissured groundwater reservoir with fracture skin, *Water Resour. Res.*, 20(7), 831, 1984.

Moench, A. F. and Hsieh, P. A., Analysis of slug test data in a well with finite-thickness skin, in *Memoirs 17th Int. Congr. Hydrogeol Rocks Low Permeability,* 17(2), Int. Assoc. Hydrogeologists, 1985a, 17.

Moench, A. F. and Hsieh, P. A., Comment on "Evaluation of slug tests in wells containing a finite-thickness skin" by C.R. Faust and J.W. Mercer, *Water Resour. Res.*, 21(9), 1459, 1985b.

Morin, R. H., LeBlanc, D. R., and Teasdale, W. E., A statistical evaluation of formation disturbance produced by well-casing installation methods, *Ground Water,* 26(2), 207, 1988.

Neuman, S. P., Analysis of pumping test data from anisotropic unconfined aquifers considering delayed gravity response, *Water Resour. Res.*, 11(2), 329, 1975.

Neuzil, C. E., On conducting the modified "slug" test in tight formations, *Water Resour. Res.*, 18(2), 439, 1982.

Nguyen, V. and Pinder, G. F., Direct calculation of aquifer parameters in slug test analysis, in *AGU Water Resour. Monogr. No. 9,* Rosenshein, J. and Bennett, G. D., Eds., 1984, 222.

Nielsen, D. M. and Schalla, R., Design and installation of groundwater monitoring wells, in *Practical Handbook of Ground-Water Monitoring,* Nielsen, D. M., Ed., Lewis Pub., Boca Raton, FL, 1991, 239.

Novakowski, K. S., Analysis of pulse interference tests, *Water Resour. Res.*, 25(11), 2377, 1989.

Novakowski, K. S., Interpretation of the transient flow rate obtained from constant-head tests conducted *in situ* in clays, *Can. Geotech. J.*, 30(4), 600, 1993.

Novakowski, K. S. and Bickerton, G.S., Measurement of the hydraulic properties of low-permeability rock in boreholes, *Water Resour. Res.*, in press, 1997.

Onur, M. and Reynolds, A. C., A new approach for constructing type curves for well test analysis, *SPE Formation Evaluation,* 3(2), 197, 1988.

Orient, J. P., Nazar, A., and Rice, R. C., Vacuum and pressure test methods for estimating hydraulic conductivity, *Ground Water Monitoring Rev.*, 7(1), 49, 1987.

Ostrowski, L. P. and Kloska, M. B., Use of pressure derivatives in analysis of slug test or DST flow period data, *SPE Paper 18595,* Soc. of Pet. Engs., 1989.

Palmer, C. D. and Paul, D. G., Problems in the interpretation of slug test data from fine-grained tills, in *Proc. NWWA FOCUS Conf. Northwestern Ground Water Issues,* National Water Well Association, 1987, 99.

Pandit, N. S. and Miner, R. F., Interpretation of slug test data; *Ground Water,* 24(6), 743, 1986.

Papadopulos, I. S. and Cooper, H. H., Jr., Drawdown in a well of large diameter, *Water Resour. Res.*, 3(1), 241, 1967.

Papadopulos, I. S., Bredehoeft, J. D., and Cooper, H. H., Jr., On the analysis of "slug test" data, *Water Resour. Res.*, 9(4), 1087, 1973.

Patterson, R. J. and Devlin, J. F., An improved method for slug tests in small-diameter piezometers, *Ground Water*, 23, 804, 1985.

Peres, A. M. M., Onur, M., and Reynolds, A. C., A new analysis procedure for determining aquifer properties from slug test data, *Water Resour. Res.*, 25(7), 1591, 1989.

Pickens, J. F., Grisak, G. E., Avis, J. D., Belanger, D. W., and Thury, M., Analysis and interpretation of borehole hydraulic tests in deep boreholes: Principles, model development, and applications, *Water Resour. Res.*, 23(7), 1341, 1987.

Priddle, M., A slug test packer for five-centimeter (two-inch) wells, *Ground Water*, 27, 713, 1989.

Prosser, D. W., A method of performing response tests on highly permeable aquifers, *Ground Water*, 19(6), 588, 1981.

Ramey, H. J., Jr., Verification of the Gladfelter-Tracey-Wilsey concept for wellbore storage-dominated transient pressures during production, *J. Can. Pet. Tech.*, 15(2), 84, 1976.

Ramey, H. J., Jr. and Agarwal, R. G., Annulus unloading rates as influenced by wellbore storage and skin, *Soc. Pet. Engr. J.*, 12(5), 453, 1972.

Ramey, H. J., Jr., Agarwal, R. G., and Martin, I., Analysis of "slug test" or DST flow period data, *J. Can. Pet. Technol.*, 14, 53, 1975.

Roscoe Moss Company, *Handbook of Ground Water Development*, John Wiley & Sons, New York, 1990.

Ross, B., Theory of the oscillating slug test in deep wells, in *Memoirs 17th Int. Congr. Hydrogeol. Rocks Low Permeability*, 17(2), Int. Assoc. Hydrogeologists, 1985, 44.

Rovey, C. W., II and Cherkauer, D. S., Scale dependency of hydraulic conductivity measurements, *Ground Water*, 33(5), 769, 1995.

Sageev, A., Slug test analysis, *Water Resour. Res.*, 22(8), 1323, 1986.

Sepulveda, N., Computer algorithm for the analysis of underdamped and overdamped water-level responses in slug tests, *U.S. Geol. Surv. Invest. Rep. 91-4162*, 1992.

Shapiro, A. M. and Greene, E. A., Interpretation of prematurely terminated air-pressurized slug tests, *Ground Water*, 33(4), 539, 1995.

Smith, J. L., Spatial variability of flow parameters in a stratified sand, *Math. Geol.*, 13(1), 1, 1981.

Spane, F. A., Jr., Applicability of slug interference tests for hydraulic characterization of unconfined aquifers. 1. Analytical assessment, *Ground Water*, 34(1), 66, 1996.

Spane, F. A., Jr. and Wurstner, S. K., DERIV: A program for calculating pressure derivatives for use in hydraulic test analysis, *Ground Water*, 31(5), 814, 1993.

Spane, F. A., Jr., Thorne, P. D., and Swanson, L. C., Applicability of slug interference tests for hydraulic characterization of unconfined aquifers: 2. Field examples, *Ground Water*, 34(5), 925, 1996.

Springer, R. K. and Gelhar, L. W., Characterization of large-scale aquifer heterogeneity in glacial outwash by analysis of slug tests with oscillatory responses, Cape Cod, Massachusetts, *U.S. Geol. Surv. Water Resour. Inv. Rep. 91-4034*, 36, 1991.

Stanford, K. L., Butler, J. J., Jr., McElwee, C. D., and Healey, J. M., A field study of slug tests in wells screened across the water table, *Kans. Geol. Surv. Open-File Rep. 96-46*, 1996.

Stone, D. B. and Clarke, G. K. C., Estimation of subglacial hydraulic properties from induced changes in basal water pressure: A theoretical framework for borehole response tests, *J. of Glaciol.*, 39(132), 327, 1993.

Streltsova, T. D., *Well Testing in Heterogeneous Formations*, John Wiley & Sons, New York, 1988.

Sudicky, E. A., A natural gradient experiment on solute transport in a sand aquifer: Spatial variability of hydraulic conductivity and its role in the dispersion process, *Water Resour. Res.*, 22(13), 2069, 1986.

Theis, C. V., The relation between the lowering of the piezometric surface and the rate and duration of discharge of a well using ground-water storage, *Trans. AGU*, 16th Ann. Mtg., pt. 2, 519, 1935.

Tongpenyai, Y. and Raghavan, R., The effect of wellbore storage and skin on interference test data, *J. Pet. Tech.*, 33(1), 151, 1981.

U.S. Dept. of Navy, Bureau of Yards and Docks, *Design Manual: Soil Mechanics, Foundations, and Earth Structures*, DM-7, Chap. 4, 1961.

Van der Kamp, G., Determining aquifer transmissivity by means of well response tests: The underdamped case, *Water Resour. Res.*, 12(1), 71, 1976.

Van Rooy, D., A note on the computerized interpretation of slug test data, *Inst. Hydrodyn. Hydraulic Eng. Prog. Rep. 66*, Tech. Univ., Denmark, 47, 1988.

Walter, G. R. and Thompson, G. M., A repeated pulse technique for determining the hydraulic properties of tight formations, *Ground Water*, 20(2), 186, 1982.

Wang, J. S. Y., Narasimhan, T. N., Tsang, C. F., and Witherspoon, P. A., Transient flow in tight fractures, in *Proc. Invitational Well-Testing Symp.*, Lawrence Berkeley Lab. Rep. LBL-7027, Berkeley, 1977, 103.

Widdowson, M. A., Molz, F. J., and Melville, J. G., An analysis technique for multilevel and partially penetrating slug test data, *Ground Water*, 28(6), 937, 1990.

Wilson, N., *Soil Water and Ground Water Sampling*, CRC Press/Lewis, Boca Raton, FL, 1995.

Wylie, A. and Magnuson, S., Spreadsheet modeling of slug tests using the Van der Kamp method, *Ground Water*, 33(2), 326, 1995.

Youngs, E. G., Shape factors for Kirkham's piezometer method for determining the hydraulic conductivity of soil *in situ* for soils overlying an impermeable floor or infinitely permeable stratum, *Soil Sci.*, 106, 235, 1968.

Zlotnik, V., Interpretation of slug and packer tests in anisotropic aquifers, *Ground Water*, 32(5), 761, 1994.

Index

D

I

K

L

M

N

O

P

Q

R

S

T

U

V

W